Genome Annotation

CHAPMAN & HALL/CRC
Mathematical and Computational Biology Series

Aims and scope:

This series aims to capture new developments and summarize what is known over the entire spectrum of mathematical and computational biology and medicine. It seeks to encourage the integration of mathematical, statistical, and computational methods into biology by publishing a broad range of textbooks, reference works, and handbooks. The titles included in the series are meant to appeal to students, researchers, and professionals in the mathematical, statistical and computational sciences, fundamental biology and bioengineering, as well as interdisciplinary researchers involved in the field. The inclusion of concrete examples and applications, and programming techniques and examples, is highly encouraged.

Series Editors

N. F. Britton
Department of Mathematical Sciences
University of Bath

Xihong Lin
Department of Biostatistics
Harvard University

Hershel M. Safer
School of Computer Science
Tel Aviv University

Maria Victoria Schneider
European Bioinformatics Institute

Mona Singh
Department of Computer Science
Princeton University

Anna Tramontano
Department of Biochemical Sciences
University of Rome La Sapienza

Proposals for the series should be submitted to one of the series editors above or directly to:
CRC Press, Taylor & Francis Group
4th, Floor, Albert House
1-4 Singer Street
London EC2A 4BQ
UK

Published Titles

Published Titles (continued)

Chapman & Hall/CRC Mathematical and Computational Biology Series

Genome Annotation

Jung Soh, Paul M.K. Gordon,
and Christoph W. Sensen

CRC Press
Taylor & Francis Group
Boca Raton London New York

CRC Press is an imprint of the
Taylor & Francis Group, an **informa** business

A CHAPMAN & HALL BOOK

CRC Press
Taylor & Francis Group
6000 Broken Sound Parkway NW, Suite 300
Boca Raton, FL 33487-2742

© 2013 by Taylor & Francis Group, LLC
CRC Press is an imprint of Taylor & Francis Group, an Informa business

No claim to original U.S. Government works

Printed on acid-free paper
Version Date: 20121128

International Standard Book Number: 978-1-4398-4117-4 (Hardback)

Library of Congress Cataloging-in-Publication Data

Soh, Jung.
  Genome annotation / Jung Soh, Paul M.K. Gordon, Christoph W. Sensen.
    pages cm. -- (Chapman & Hall/CRC mathematical and computational biology series)
  "A CRC title."
  Includes bibliographical references and index.
  ISBN 978-1-4398-4117-4 (hardcover : alk. paper)
  1. Bioinformatics. 2. Genomics--Data processing. 3. Human genome. I. Gordon, Paul
M. K. (Paul-Michael Kempton), 1976- II. Sensen, C. W. (Christoph W.) III. Title.

QH447.S634 2013
572'.330285--dc23   2012026279

Visit the Taylor & Francis Web site at
http://www.taylorandfrancis.com

and the CRC Press Web site at
http://www.crcpress.com

Contents

Preface

The year 1995 saw the arrival of the first completed microbial genomes, *Haemophilus influenzae* and *Mycoplasma genitalium*. Several years of struggle for a complete genome was ended by the group at The Institute for Genomic Research (TIGR). From today's point of view, 16 years and more than a thousand completed genomes later (including, of course, the human genome), it may be hard to understand how much of an accomplishment this was. At the time, however, most laboratories around the world were still sequencing on slab gels using radioisotopes as the label for the fragments, which represent the DNA sequence.

When we sit down to "browse" genomes today, we do not often remember the days before e-mail and the Internet, or the days before automated DNA sequencing became a commodity. But it is certainly worthwhile to take a look back, as many of the design decisions that were made in the last 16 years influence the way we deal with genomic information today.

When the first genomes were presented at a by-invitation-only meeting in Worcester, England, in 1995, the only tool that was capable of handling such a large file was a word processor. Therefore, the sequence was first presented to the scientists at the meeting as a character file, which was scrolling on the screen behind the speaker. One of the major problems with handling a large DNA sequence file at the time was that most bioinformatics software was only tailored for DNA fragments of a size much less than a complete microbial genome, typically no more than approximately 100 kilobase pairs. The first automated genome analysis and annotation systems were barely emerging in 1995, and thus the handling of a complete genome with a size of more than a million base pairs all at once was impossible.

The Web was a fledgling entity in 1995, with not much power and entirely based on the Hypertext Markup Language (HTML). It became clear very quickly that only large communities of scientists with a diverse background could really make sense of the genomic information, provided that they were enabled to collaborate, and thus the Web quickly became the vehicle by which genome annotations were created and exchanged among scientists. The first automated genome analysis and annotation systems, which were Web-based, initially produced tabular output that listed the location of potential genes and gene functions, which were predicted mostly by database comparison. It became obvious very early that this was not sufficient for biologists, therefore graphical subsystems were added, which are today part and parcel of all genome analysis and annotation systems and are probably the only part of an automated genome analysis and annotation system that most users ever encounter.

Over time, the Web developed into the massive entity it is today, with many additions to the Web technologies, which were utilized in turn by the developers of today's genome analysis and annotation tools. The three most useful tools in this context were probably (1) the creation of the programming language Java by James Gosling, which allowed the development of truly platform-independent applications; (2) the introduction of Extensible Markup Language (XML), which could adequately be used for the description of biological and medical objects; and (3) the creation of Web services (for example, the BioMOBY system), which made distributed computing simple and easy, and allowed the transparent and seamless integration of new bioinformatics tools into Web-based bioinformatics applications.

DNA sequencing technology has progressed in several iterations to today's level, which is called "next-generation sequencing," but really represents the third or fourth generation of DNA sequencing technologies, with yet another generation just around the corner. The sheer amount of DNA sequence, which can be produced on a single device today, is mind-boggling. It has literally become possible to resequence genomes the size of the human genome within a few hours in a single laboratory and the $1000 human genome is on the horizon. At the time of this writing (2012), very few genome annotation pipelines are capable of dealing with this information volume and new strategies need to be developed to accommodate the needs of today's genome researchers.

In the near future, everyone will be able to carry their genome sequence on some kind of data storage device and diagnostics might become largely based on the results of genomics screens, which will be cheaper than today's advanced imaging technologies (MRI and CT scans, for example). Thoroughly annotated genomic information and the integration of all information into a single model will be a prerequisite to successful approaches to individualized medicine, the development of advanced crops and the sustainable production of food, medical research and development, and the development of new and sustainable energy sources.

This book attempts to introduce the topic of automated genome analysis and annotation. The initial chapters take the reader through the last 16 years, explaining how the current analysis strategies were developed. This is followed by the introduction of up-to-date tools, which represent today's state of the art. The authors also discuss strategies for the analysis and annotation of next-generation DNA sequencing data. This book is intended to be used by professionals and students interested in entering the field.

We would like to thank Hershel Safer and the editorial team at CRC Press/Taylor & Francis for their patience while creating this book.

Authors

Jung Soh is a research associate at the University of Calgary, in Alberta, Canada. He was born in Seoul, Korea, and received a Ph.D. in computer science from the University at Buffalo, State University of New York, where he worked at the Center of Excellence for Document Analysis and Recognition (CEDAR). From 1992 to 2004, he worked as a principal research scientist at the Electronics and Telecommunications Research Institute (ETRI), Daejeon, Korea, in areas such as biometrics, human–computer interaction, robotics, video analysis, pattern recognition, and document analysis. Dr. Soh moved to Calgary in 2005. His current research interests include bioinformatics, machine learning, and biomedical data visualization.

Paul Gordon is the bioinformatics support specialist for the Alberta Children's Hospital Research Institute at the University of Calgary. He was born in Halifax, Nova Scotia, Canada. He received his bachelor of science (first class honors) and master's of computer science from Dalhousie University (Halifax). He worked at the Canadian National Research Council's Institute for Information Technology and at its Institute for Marine Biosciences

(1996–2001) before moving to Calgary. From 2002–2011, he was a lead programmer on several genome Canada-funded projects, supported by the University of Calgary. In the new era of the $1000 genome, Gordon's current work focuses on developing bioinformatics techniques for personalized medicine.

Christoph W. Sensen is a professor of bioinformatics at the University of Calgary, in Alberta, Canada. He was born in Oberhausen-Sterkrade, Germany, and studied biology in Mainz, Düsseldorf, and Köln. From October 1992 to November 1993, he worked as a visiting scientist at the European Molecular Biology Laboratory (EMBL) in Heidelberg, Germany. At the beginning of 1994, he moved to Canada. Until his move to Calgary in February 2001, he worked as a research officer at the National Research Council of Canada's Institute for Marine Biosciences (NRC-IMB). His research interests include genome research and bioinformatics.

Contributor

Stephen Strain was born in Szeged, Hungary. He studied visual arts and music in Debrecen, Szeged, and Budapest. In 1981, he emigrated to Calgary, Alberta, Canada, where he lives and works as an artist. He is known for his interpretation of flamenco music (www.flamencoelegante.com) and as a painter.

His painting *Magpie* is used as the cover design.

CHAPTER 1

DNA Sequencing Strategies

1.1 THE EVOLUTION OF DNA SEQUENCING TECHNOLOGIES

Methods for DNA sequencing were invented multiple times, first by chemical means (Maxam and Gilbert, 1977) and later using biochemical approaches (Sanger and Coulson, 1975). Initially, radioactive compounds were used to make the DNA bands, which represent base pairs (bp), detectable by autoradiography after an electrophoretic separation. A major step forward was the switch to fluorescent labels in the 1980s, which could be detected during the electrophoretic separation and automatically recorded by computers, instead of humans reading autoradiograms, which were then manually converted to computer files. Naturally, the manual editing necessary for the creation of the early DNA sequence files led to a high error rate, which should be kept in mind when comparing DNA sequences downloaded from public repositories, such as GenBank.

The early automated DNA sequencing machines allowed for the production of a few kilobase pairs of DNA sequence per day and machine. Today, this has increased tremendously through the invention of the so-called next-generation DNA sequencing technologies and the integration of robotic workstations into the DNA sequencing workflow. It is now possible to produce millions of reads per day and device. Still the read length obtained in 2012, with few exceptions, does not yet reach that of the first-generation automated DNA sequencing machines, which peaked at about 1200 bp. The Roche 454-type systems (www.454.com), which are currently the mainstream long-read machines, generate an average read length of around 450 bp at the most. Finally, this limit is about to be broken by yet another generation of DNA sequencing devices, which are expected to produce read lengths of at least several kilobase pairs per read. The new system from Pacific Biosciences, called PacBio RS (www.pacificbiosciences.com), can already reach this level of longer read lengths but currently with major trade-offs in sequence accuracy.

Regardless of the DNA sequencing technology used for automated DNA sequencing, all approaches ultimately lead to the creation of a so-called trace file, which captures essentially a sequence of base pairs, which never represents more than a fraction of a genome. The trace files are essentially graphical representations of the DNA sequencing progress, either represented as a succession of DNA fragments throughout a separation by size or the representation of a DNA synthesis on a chip or in a flow cell over time. The DNA sequence can be extracted from the trace file as a character file. Typically, the character file is formatted as a "FASTA"-format file, which essentially contains the DNA sequence (usually 60 characters per line) and a single description line, which is preceded by a ">" character on the first line. Multiple sequence files can be combined into a single file with multiple pairs of a description line and a sequence, each corresponding to one sequence and preceded by a ">" character, which leads to a so-called multiple FASTA file.

1.2 DNA SEQUENCE ASSEMBLY STRATEGIES

Until now, there is no DNA sequencing technology that directly leads to a complete genome with a single sequencing run. In all cases, the complete contiguous sequences (i.e., contigs) has to be constructed from small individual sequence reads, which range from 30 to a few hundred base pairs. The necessary sequence assembly can be done using three fundamental genome assembly strategies: primer walking, shotgun assembly, or a mixed strategy.

Initially, much of the DNA sequence production was done by so-called primer walking (Voss et al., 1993). This essentially meant that after each sequence run, once the new DNA sequences were integrated into the assembly project, a set of oligonucleotides was calculated (to obtain the next DNA sequencing primers), which could extend the existing contigs. This approach was quite slow, as the new primers needed to be synthesized before every new sequencing run and could fail for a number of reasons, and it was also expensive, as dedicated primers needed to be synthesized for each individual sequencing run. First and foremost, problems arose from repetitive elements within the clone that was being sequenced, which were not yet known to the researcher and therefore not taken into account during the primer calculation, and resulted in DNA sequencing errors.

This was especially true for single-stranded DNA sequences, where the usual sequencing errors led to the calculation of primers that were not priming correctly due to built-in mistakes. An advantage of the primer-walking strategy was the low coverage necessary, even when both DNA strands were sequenced completely. In many cases, the final coverage was between three and four times the genome equivalent. Figure 1.1 shows an overview of the primer-walking sequencing strategy.

Craig Venter, then at The Institute for Genomic Research (TIGR), and his team can be credited for the invention of the "shotgun sequencing" approach (Venter et al., 1996). In the early version of this DNA sequencing strategy, which was still based on cloned DNA fragments, only universal primer pairs were used to sequence into the cloned DNA inserts with much higher redundancy (6- to 10-fold genome coverage) than in the primer-walking approach. Once enough end sequences were obtained, they were assembled into contigs. Figure 1.2 shows an overview of the shotgun sequencing strategy.

If all regions of a genome would be equally clonable, this strategy could have rapidly yielded complete genome sequences as only standard components, such as universal primers, were required during sequence production, but this was rarely ever the case. Therefore, applying this strategy in its pure form led to almost completely sequenced genomes, but in the case of microbial genomes as an example, many gaps remained, often numbering several hundred gaps per 2 to 3 mega bp of genomic sequence. In addition, in most cases many regions within the genome were only

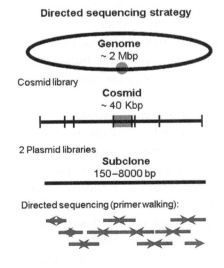

FIGURE 1.1   Overview of the primer-walking DNA sequencing strategy.

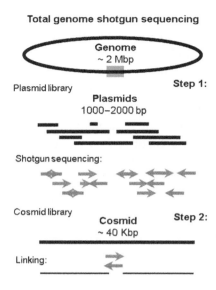

FIGURE 1.2   Overview of the shotgun DNA sequencing strategy.

sequenced on a single DNA strand (most likely also due to cloning arti-facts). This increased the error rate in the finished genome, as multiple instances of a single base occurring in sequence are generally best resolved by the sequencing of both DNA strands. The redundancy of a shotgun-sequenced genome is typically much higher than that of one sequenced by primer walking, with 8- to 10-fold, but sometimes more than 20-fold being reported for the finished product.

Ultimately, almost all groups attempting to sequence complete genomes settled on a mixed strategy, where the bulk of the genome sequencing was accomplished through shotgun sequencing and the gaps were closed through primer walking on a set of large-insert clones. Figure 1.3 shows an overview of the mixed sequencing strategy.

This approach yields complete genomic sequences at a much lower cost than the genomes finished entirely by primer walking, as only very few dedicated DNA sequencing primers are required with almost the same ultimate DNA sequence accuracy. Typically, in the case of a microbial genome with a size of 2 to 3 mega bp, between 200 and up to 500 gaps need to be closed by primer walking, most of these being fairly small. In support of gap closure, primer-walking technology using large-insert libraries, such as cosmids, fosmids, or lambda clones were developed, which can be used to efficiently complement the shotgun sequencing results. Today,

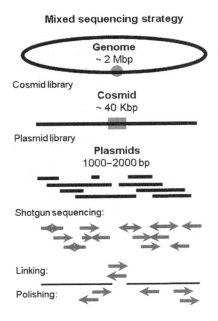

FIGURE 1.3 Overview of the mixed DNA sequencing strategy.

most of this is already history, as next-generation DNA sequencing is replacing most of the above. Still the aforementioned techniques have led to the generation of more than a thousand genomes, many of them of very good quality (1 error per 10,000 bp or less). These early complete genomes provide the templates for today's sequencing experiments.

1.3 NEXT-GENERATION SEQUENCING

The latest DNA sequencing technologies certainly influence the current DNA sequencing strategy and lead to new approaches in biochemistry and biotechnology. Two major types of next-generation DNA sequencing technology are being developed: short-read systems, with read lengths up to approximately 100 bp per DNA fragment, and long-read systems, with average read lengths of at least 300 bp and potential read lengths of several kilobase pairs per fragment. There are many different technologies that are currently being explored for next-generation sequencing, with the Illumina, ABI SOLiD, and Roche 454 sequencing approaches being in the lead at present. The latest sequencing technologies do not require cloned DNA as a prerequisite, thus sequencing can be accelerated tremendously. A single machine run can yield several million reads, which is sufficient for the shotgun assembly of a microbial genome.

As several hundred genomes, including the human genome, have been completely sequenced to date and thousands more have been completed with various grades of quality, it is now often of little interest to complete genomes entirely. As the cost for DNA sequence production has been lowered considerably, when compared to the older capillary DNA sequencing technology (a human genome equivalent can be generated for less than $10,000 in 2012 and might be possible to be generated for less than $1,000 shortly), sequence redundancy can be increased considerably to factors of 30- to 40-fold or even higher coverages. This often leaves only 20 to 50 gaps in the genomic sequence of a microbial genome, which might or might not be closed through traditional primer-walking strategies. Often, the new genomic sequence is close enough in similarity to an existing complete genome that the fully completed genome can be used as a "scaffold" to align the newly generated sequence in a meaningful way. The low number of gaps means that in a microbial genome less than 1% of the genes are not fully characterized. This is sufficient for almost all genome research needs.

New analysis strategies are being developed, which supersede many older molecular biological techniques, including PCR experiments, DNA mapping technologies, and even gene chips. Complete sequencing of the "total genome" of an organism is now possible. In the human case, this means not only the generation of the DNA sequence of the human chromosomes, but also the sequencing of all DNA-containing organisms and viruses in the body fluids, such as blood and also those in the gut. The total DNA content of a human being is at least a thousand times larger than the DNA content of the chromosomes, posing major challenges for the annotation of this entire conglomerate of DNA molecules.

1.4 SEQUENCING BIAS AND ERROR RATES

In our opinion, to date, there is really no "complete and error free" genome sequence that was ever characterized by humans. Many sources of error exist, leading to certain error rates in all DNA sequences, which have been submitted to the public repositories. Typically, the generation of a new sequence involves a large number of individual steps, from the isolation of the original DNA, to the preparation for DNA sequencing using various kits, to the actual operation of the DNA sequencing equipment, and finally to the assembly and annotation. In most steps, direct human involvement is still necessary. Errors are manifold in nature, from single-base pair differences (Meacham et al., 2011), which can lead to frameshifts in the sequence, to different assemblies of the same sequence which lead to a false representation of the genomic arrangement and gene order (for example, two assemblies of the bovine genome currently exist, which are based on the same data, but differ considerably) (Florea et al., 2011), to "missing" genes or regions based on the erroneous assembly of repetitive regions (for example, in the human genome) (Semple et al., 2002) in a genome. Thus already the input into the genome annotation process is not without flaws. The goal should be to use the best possible input sequence. This can be best achieved by using at least two different sequencing techniques simultaneously while generating the final sequence. This also helps to iron out the bias, which is introduced by every sequencing technology. Some machines cannot deal with high-GC (guanine–cytosine) content DNA fragments the same way they deal with low-GC content fragments, some machines balk at homopolymer stretches in the DNA sequence, and some machines lose accuracy during the sequencing process to a large degree, calling the final bases in the sequencing run with no more than 70% accuracy.

In the future, the combination of short reads (around 100 bp), with reads generated at both ends of the DNA fragment (paired-end sequencing) and very long reads (over 1000 bp) will probably prevail for *de novo* sequencing, while short reads alone will be used if a more or less complete genome template already exists (for example, in the case of the human genome).

REFERENCES

Florea, L., Souvorov, A., Kalbfleisch, T.S., Salzberg, S.L. 2011. Genome assembly has a major impact on gene content: A comparison of annotation in two *Bos taurus* assemblies. *PLoS One* 6:e21400.

Maxam, A.M., Gilbert, W. 1977. A new method for sequencing DNA. *Proc. Natl. Acad. Sci. USA* 74(2):560–564.

Meacham, F., Boffelli, D., Dhahbi, J., Martin, D.I., Singer, M., Pachter, L. 2011. Identification and correction of systematic error in high-throughput sequence data. *BMC Bioinformatics* 21:451.

Sanger, F., Coulson, A.R. 1975. A rapid method for determining sequences in DNA by primed synthesis with DNA polymerase. *J. Mol. Biol.* 94(3):441–448.

Semple, C.A., Morris, S.W., Porteous, D.J., Evans, K.L. 2002. Computational comparison of human genomic sequence assemblies for a region of chromosome 4. *Genome Res.* 12:424–429.

Venter, J.C., Smith, H.O., Hood, L. 1996. A new strategy for genome sequencing. *Nature* 381(6581):364–366.

Voss, H., Wiemann, S., Grothues, D., et al. 1993. Automated low-redundancy large-scale DNA sequencing by primer walking. *Biotechniques* 15:714–721.

CHAPTER 2

Coding Sequence Prediction

2.1 INTRODUCTION

Gene structure prediction is the primary task in genome annotations, therefore much work has been put into developing methods to accurately make these predictions. Methods for the prediction of protein coding sequence (CDS) are tremendously varied, and their suitability depends on many factors such as biophysical parameters, for instance the guanine–cytosine (G+C) content of the DNA, the phylogenetic position of the organism, the types and quality of sequence available, and the number of novel genes expected. In both prokaryotes and eukaryotes, the three broad approaches to finding coding sequence are transcript mapping, statistical modeling of gene structure, and homology modeling. In eukaryotes, an additional consideration is the prediction of alternative splicing instances for the predicted CDSs. Each approach is addressed here from a perspective of the algorithmic design or sequence feature in order to provide knowledge applicable to many types of genomics data. But the reader is encouraged to explore the citations of the original literature for details of how extant software implement these methods.

2.2 MAPPING MESSENGER RNA (mRNA)

Since CDSs are defined by the proteins they encode and the proteins are translated from messenger RNAs (mRNAs), mRNA transcripts are considered the gold standard for gene sequence elucidation. In the past this mostly involved Sanger sequencing of cDNA generated from mRNA, so-called expressed sequence tags (ESTs). This has largely been supplanted by the use of 454 (pyro-) sequencing. Sequence data generated by 454 sequencing can be assembled using similar assembly algorithms to Sanger sequences, since both are usually of similar length (between 300 and 500 bp). Due to the sheer volume though, specialized assembly software packages exist that better handle the large volume of reads involved. Popular packages include the traditional Phrap (www.phrap.org) and CAP3 (Huang and Madan, 1999), the high-volume MIRA (Chevreux et al., 2004) and SeqMan (www.dnastar.com) assembly packages, as well as the assembler provided by the manufacturer of the 454 machine (www.454.com), Newbler. Another commercial assembler, known for its tremendous speed, is the CLC Bio assembler (www.clcdenovo.com).

More recently, the ease of sample preparation and low cost of short-read transcript sequencing (e.g., Illumina RNA-Seq) have made it possible to effectively reveal the transcriptome of any organism. The depth of coverage obtained through RNA-Seq data has greatly enhanced prokaryotic and eukaryotic transcriptomics, demonstrating that loci are differentially expressed and in the case of eukaryotes can be alternatively spliced far beyond what was known from the results obtained by the generation of ESTs. Short reads can be assembled into transcripts using computationally intensive software such as ALLPATHs (Butler et al., 2008), PEACE (Rao et al., 2010), Trans-ABySS (Robertson et al., 2010), or SOAPdenovo (soap.genomics.org.cn) and Velvet/Oases (Zerbino and Birney, 2008) assembly packages. Although some of these packages were originally designed for genomic assembly, they have been enhanced to deal with the variable read coverage of transcriptomes (which is generally not an issue in shotgun genome sequencing), as described next.

The process of assembling short transcripts into accurate full-length coding sequences is not yet a standardized process. Variation in read coverage along a transcript due to various factors, and large variation in transcript abundance mean that most *de novo* assemblies require trial and error and the combining of multiple assembly runs, using various parameter values to capture both abundant and low-coverage genes at a reasonable computational cost. The value provided by doing *de novo* assembly is that it can even be done in the absence of a genome sequence and that splice variants can be automatically identified. Boundaries of exons become clear if the transcript consensus can be mapped back to the genomic sequence.

The list of genome/transcriptome assemblers is continually growing, therefore it is best to refer to on-line forums to learn about the latest tools. One of the most popular forums is SEQanswers.com. Combined with trawling the mailing list archives for the tools, most of the tool suitability and parameterization questions can be rapidly answered.

Once contiguous sequences representing mRNAs have been assembled, they can be aligned back to an already existing reference genome. The intron–exon structure of the genes can then be revealed through the use of a double-affine, also known as intron-tolerant, alignment algorithm (Garber et al., 2011). Double-affine searches have a high penalty for small gaps (insertions, deletions, or sequencing errors) and a smaller one for large gaps (introns). This promotes the correct alignment of cDNA sections.

Aligning across introns can lead to slight misalignments. For example, given an alignment of the form:

```
CATCCTTGATACGACGCTGCTGGTGGAGAATGCGATCCACTAGAACGAACATCAGGAATC

||||||||||||||||||||||||

CATCCTTGATACGACGCTGCT----------------------------------------


ACATTCAAAAAAAAAAAAAAAAAAAAAAAAAGAAGAAGAATTACAGCAAGAAAGCGGTATGC

                                            ||||||||||||||||||

-----------------------------------------------GCAAGAAAGCGGTATGC
```

The read could have just as easily been aligned with the G at the start of the second exon moved to the end of the first exon without disturbing the protein translation. This would have maintained the splice acceptor AG that is almost universal in the species being studied. Good software should take preferred splice sites for a species into account in order to provide accurate exon alignment.

Alternatively, the short reads can be mapped back to the genome using a fast short-read alignment tool. While many such alignment tools exist, of particular note are those that are consistently updated and support the popular SAM (Sequence Alignment/Map) file format for read mapping, especially BFAST (Homer et al., 2009), BowTie (Langmead et al., 2009), and MosaikAligner (http://code.google.com/p/mosaik-aligner). A special subcategory of short-read alignment tools, important for mapping to multiexon genes, are gapped alignment tools such as the Burrows-Wheeler Aligner (Li and Durbin, 2010), GEM split-mapper (http://gemlibrary. sf.net), Karma (http://genome.sph.umich.edu/wiki/Karma), and TopHat

(Trapnell et al., 2009). Intron-spanning read mappers are a relatively new category, and the number of mapped reads can vary considerably from one program to the next. These programs also tend to be quite sensitive to the input parameters used, and generally a trade-off between speed and accuracy exists. Established mRNA-to-genome mappers such as BLAT (Kent, 2002) can also be used, as long as the data set is small enough or processing time is not a critical factor. Most modern tools that incorporate short-read data into gene predictions and visualizations expect SAM formatted files as the input, therefore the need for multigigabyte-file-format conversion for these other tools must be taken into consideration as well.

The mapping of reads directly back to the genome is relatively straightforward, with a few parameters such as the number of mismatches allowed and whether the quality values of the bases are taken into consideration. The exceptions to this are intron-spanning reads and multicopy genes. Software tools that perform this type of mapping are quite sensitive to the parameters given. The length of the hit "anchor" should be somewhat less than half the length of the reads being mapped in order to catch reads spanning introns. One must specify parameters such as minimum and maximum intron size and whether noncanonical splice sites are known to occur. Thorough background knowledge about what is biologically viable for the organism being studied is essential to maximize the accuracy and relevance of the spliced-read predictions.

Read mapping is also highly parallelizable, therefore several short-read alignment tools have been adapted to work on computer clusters, and so-called compute clouds, where processor time can be rented by the hour. The growing adoption of cloud computing means that even individual researchers who do not own much infrastructure themselves can perform mRNA analysis. As their names suggest, tools such as CloudBurst (Schatz, 2009) for mapping and Contrail (http://contrail-bio.sf.net) for *de novo* assembly have been built specifically to use cloud computing. In other cases, tools that use the OpenMPI parallelization libraries can seamlessly transition to the Amazon EC2 cloud, as has been reported for ABySS. Some knowledge of the resource requirements for the types of assemblies being performed is required; several tools can use more than 30GB of RAM, and require the rental of time on suitably equipped machines.

Although some short-read mapping tools and gene prediction tools have been mentioned here, this is an evolving field with many new tools being currently produced. Benchmarking of such tools is taking place through the rGASP series of competitions (www.sanger.ac.uk/PostGenomics/

encode/RGASP.html). The results of this competition provide a good reference point for the relative strengths of each program.

Although mapping of short reads or assembled mRNAs back to the genome can be very helpful in determining correct intron and exon boundaries, it is only one component of the overall gene prediction process. This is because (1) it is unlikely that full-length mRNA sequences can be obtained for all genes; (2) in some species transcripts may encode operons of more than one gene; and (3) start and stop codons cannot be determined from the mRNA alone due to multiple possible translation frames and factors affecting translation initiation. As such, transcript information is therefore supplemented in most gene prediction software by statistical models of gene feature sequences.

2.3 STATISTICAL MODELS

Most *ab initio* gene model predictors are based on a data structure known as a generalized hidden Markov model (GHMM). Conceptually, it can be thought of as a graph where the nodes represent events in the gene's expression, and arrows connecting the nodes represent the scanning of the sequence, as shown in Figure 2.1. It should be noted that the actual GHMMs used in prediction programs have many more states (nodes) and transitions (arrows) than in this example.

Each gene model is calculated by progressively scanning each base of the genomic sequence, looking for signals to transition from one state to the next. In the case of a transcription start, the signal being searched for is not necessarily the exact string GCCCTA but perhaps a pattern such as (G/C) CCXTA. (Note: G, guanine; C, cytosine; T, thymine; A, adenine; X, any

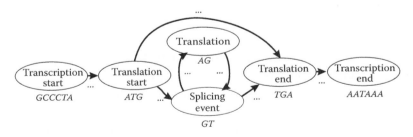

FIGURE 2.1 A simple generalized hidden Markov model, as used by gene prediction algorithms. DNA sequence is consumed by the first node of the model until an expected sequence is found, then the next node in the graph consumes the sequence, and this process continues. A gene's boundaries and features are defined by these transition points.

single nucleotide.) The configuration of these search patterns is achieved by providing a training set of known transcription start sites to the prediction tool. Likewise, the real start codon is not necessarily the first ATG encountered, so a training set of known full-length coding sequences can greatly assist in building a model of what the DNA sequence *around* the real start codons looks like. The sequence between the transcription start and the translation start may also have some statistical characteristics, as may the sequence between the GT and AG in an intron. The probabilities derived from the training data sets are what determine the accuracy of the subsequent predictions.

Gene features are often referred to by their acronyms in gene prediction systems. Therefore it is worthwhile remembering these as they are introduced here. Statistical models may include the features described next, roughly in genomic order.

2.3.1 5′ Untranslated Region

The 5′ untranslated region (5′ UTR) is the region between the transcription start site and the translation start codon. Many eukaryotic gene predictors can take advantage of user-identified 5′ UTR sequences to train their statistical models. Where sufficient aligned mRNAs or mapped short reads are available, and the start codon is known via protein homology, the identification of 5′ UTR can be based on empirical evidence for training purposes. The 5′ UTR region contains several translation signals, as detailed next. The identification of transcription-factor binding sites is a complex issue. In general the transcription-factor binding site locations are not often used to assist in the prediction of gene structures. Their discovery and potential usage is discussed later in Chapter 3.

2.3.2 −35 Signal

The −35 signal site is located approximately 35 base pairs upstream of the start codon in prokaryotes. It can be used to aid in identification of the proper binding site for the initiation complex during transcription. This signal has a consensus sequence of TTGACA (Rosenberg and Court, 1979), but is relatively weak compared to the other prokaryotic UTR signals, therefore multiple candidates often exist for a single gene.

2.3.3 B Recognition Element

The B recognition element (BRE) is found in many archaeal and eukaryotic genes (Qureshi and Jackson, 1998), just upstream of the TATA box, and consists of seven nucleotides. The first two nucleotides of the BRE sequence are either G or C, followed by either G or A. The last four nucleotides are CGCC. Transcription factor IIB (TFIIB) binds in the major groove of the DNA at the BRE.

2.3.4 TATA Box

In prokaryotes, the TATA box is known as the Pribnow box, or the −10 signal, because it is usually centered 10 bases upstream of the start codon. Although there is a TATAAT consensus (Pribnow, 1975), it is not universal. More important is that the region is rich in weak adenosine and thionine bonds. Together with the −35 signal, these patterns specify suitability for transcription initiation. It is quite common to observe multiple neighboring candidate start initiation sites, due to the short length and variability of the −10 and −35 signals, therefore the predicted patterns and their putative locations are mostly used to support rather than define the start of the coding sequence.

In eukaryotes, the TATA box is only one of many possible initiation factors. Core promoters, such as the initiator element (Inr), the GC box, or the CAAT box can replace or augment the eukaryotic TATA box. Other initiator elements specific to particular genes and tissue types exist but are not generally used for statistical prediction of genes due to their rarity.

2.3.5 Ribosomal Binding Site

The Shine–Dalgarno sequence (Shine and Dalgarno, 1975) represents the ribosomal binding site (RBS) in prokaryotes and hence the start of translation. It is usually situated within eight base pairs of the start codon. It is species specific, determined by the reverse complement of the 16S ribosomal RNA's 3′ end, and may contain mismatches (Schurr et al., 1993). Not every gene necessarily starts with an RBS, but each transcript should. This means that scores for genes in the middle of an operon (also known as polycistronic transcripts) should not be penalized for not having one.

In eukaryotes, the ribosomal binding site is referred to as the Kozak sequence (Kozak, 1986). This site does not have as strong a consensus as in prokaryotes (is not derived from the 18S rRNA), and it overlaps rather than precedes the start codon. Because of this overlap, the start codon begins at position +1, and the immediate 5′ base is position −1. The consensus sequence varies from species to species; therefore proper detection requires a training data set.

2.3.6 Start Codon

In eukaryotes, this sequence is almost universally ATG. In prokaryotes, ATG is in most instances still the dominant start codon (approximately 80%), but GTG and TTG can also be used in descending order of frequency. Although there are exceptions or substitutions to these start codons in some species, they are so infrequent that no automated gene prediction system considers them *a priori*. Evidence for alternative start codons usually comes from mass spectrometry data or *de novo* peptide sequencing and has to be entered manually when the genome analysis pipeline is being configured.

2.3.7 Protein Coding Sequence

The determination of the protein coding sequence (CDS) is of course at the heart of every gene-prediction system. Because the sequence is constrained by the codon-usage table and the protein to be produced, strong signals exist that can be predictive of CDS regions in a genome. Different feature sets of the CDS are used in the statistical models underlying various prediction programs, but they roughly fall into two categories: sequence base composition and codon usage bias. Sequence base composition in many prokaryotes can be used to predict the coding strand in the genome, where the strand containing greater than 50% A+G is usually the protein-coding instance (to our knowledge only with the exception of retroviruses). Unusual G+C composition can be used to filter genes, which might be horizontally transferred from another species, and transposable elements. None of these sequences will follow the general statistical distributions for the regular coding sequences in a given species.

For eukaryotes, it is also suggested by some programs to separate the coding sequence sets into bins, based on their G+C content. The reason is that sequences in these different bins can show subtle differences in their other sequence statistics, leading to systematic underreporting of some CDSs, if not explicitly modeled. These percentage G+C bins (called isochores) can be defined manually or automatically by some programs, such as IsoFinder (Oliver et al., 2004).

Even without comparison to databases of known proteins, it is possible to some degree to predict the coding sequence based on the low occurrence of rare codons. This is particularly useful for the determination of the correct start codon for the CDS, which is species specific. The codon adaptation index (Sharp and Li, 1987) is another statistical measure of codon usage that captures the fact that the codon used to encode an amino acid is highly correlated with the codon immediately to its 5′ end. For reasons that are not completely understood, certain synonymous pairs of codons are more common than others, and frequency is inversely correlated with translation efficiency.

Given a training set of well-characterized and ideally confirmed coding sequences, gene prediction tools can be used to choose in new sequence the reading frame that best fits the training sequence composition and codon statistical models. It is important that, for prokaryotes, the training gene set comes from throughout the genome, since a strong GC skew [(C+G)/(C−G)] usually exists, which corresponds with the origin(s) of replication.

With any statistical model, false-positive and false-negative hits may occur. The more representative the training set, the better the results become. Additionally the number of pseudogenes falsely being identified as CDSs can be quite high. It is for these reasons that usually multiple signal features are weighted and used to make the predictions, rather than relying on sequence statistics alone.

2.3.8 Donor Splice Site

The canonical intron model starts with GT and ends with AG. Depending on the organism, 1% to 2% of the splice sites may vary from this model, typically showing the sequence GC–AG or more rarely AT–AC (Clark and Thanaraj, 2002). Most software packages assume that the intron donor is GT, and therefore will not report noncanonical splice sites. An alternative, or so-called cryptic splice donor site, can be identified by mapping short reads or mRNA sequences to the genomic sequence. Compositional bias immediately surrounding the donor and acceptor sites occurs (Hertel, 2008). Gene prediction software may take this into account, based on the initial gene model training data set provided.

2.3.9 Intron Sequence

The region upstream of the intron acceptor is typically CT-rich (the so-called polypyrimidine tract), and this is preceded by an adenine (the intron's branch point) somewhere 20 to 50 bases before the acceptor. In general, most of the intron sequences will not have a base composition or codon usage that fits the CDS model. The correct identification of introns in the training data set for CDSs is critical for the correct intron prediction. In the case of retained introns, or cryptic splice sites, the compositions of the intron may be atypical, reducing the likelihood that it is automatically detected using codon usage or composition analysis alone.

2.3.10 Acceptor Splice Site

The acceptor site for introns is almost invariably AG (with the rare AC exception noted earlier), and cryptic splice acceptor sites are roughly as common as cryptic donors. Consensus in the exon immediately following the acceptor is less common than at the donor site.

2.3.11 Stop Codon

The three "nonsense" or stop codons are TGA, TAG, and TAA. This signal is very straightforward to detect once the reading frame is established. A rare exception to this is programmatic frameshifting in the CDS (generally occurring in polyT areas), which must be detected with frameshift-compensating software such as ESTScan (Iseli et al., 1999). Another exception occurs in mitochondria, where TGA is used as a codon for tryptophan. Annotation of mitochondrial and chloroplast genomes must also consider that RNA editing (generally C→U) may occur, causing a disparity between the standard genomic sequence and the actual mRNA being translated. There is growing evidence from the 1000 Genomes Project (1000 Genomes Project Consortium, 2010) and others that RNA editing is more widespread than originally thought across the eukaryotes, and not only in organelle genomes as previously thought. As no software tools currently exist that can be used to predict RNA-edited bases in eukaryotes, such determinations must be made using a combination of genomic and transcript data.

2.3.12 3′ Untranslated Region

Many gene prediction tools can be trained on 3′ untranslated region (3′ UTR) data sets to improve their accuracy, especially by avoiding the creation of "chimeric" gene predictions that merge neighboring genes into a single gene model. The 3′ UTR sequence is definitely not random, nor should it be treated as an intron or intergenic region. Besides containing terminator signals, in eukaryotes the 3′ UTR is known to be important as a target for RNA silencing. Different 3′ UTRs may also direct subcellular localization of the protein, even when the translation is identical (An et al., 2008).

2.3.13 Terminator

In many prokaryotes, the primary mechanism for intrinsic transcription termination is the presence of a G+C-rich sequence, which forms a stem with a small loop (also known as a hairpin) as single-stranded RNA.

Programs such as TransTerm (Jacobs et al., 2009) and GeSTer (Unniraman et al., 2002) can be used to automatically identify these structures, with the latter more sensitive to divergent structure and position of the terminator. The other prevalent termination method is called rho dependent because of the protein involved in facilitating termination. Rho appears to bind C-rich sequence with little secondary structure, but otherwise no consensus is known.

Three different RNA polymerases (I, II, and III) are known to function in eukaryotes. Although no specific sequence is associated with termination for the first two polymerase types, transcripts generated by RNA polymerase III are terminated in a manner similar to that of prokaryotes. Four or more T's in a GC-rich region are required as the termination signal, but the region does not need to form a stem-loop structure.

Most mRNAs in eukaryotes are polyadenylated, with notable exceptions, such as the histone genes. In this process the real 3′ end of the transcript is replaced with 20 or more adenosines. This replacement site has a consensus of AATAAA with some 3′ variation (Beaudoing et al., 2000) and can be used to boost the confidence that a prediction tool has correctly identified the last exon of the gene.

2.4 CROSS-SPECIES METHODS

The determination of gene structure through validation using existing genes and genomes falls somewhere between transcript modeling (empirical) and *ab initio* modeling (statistical). Homology modeling requires both empirical data, in the form of existing homologous data sets, and statistical data, in the form of phylogenetic inference. Cross-species validation can be based on either nucleotide sequence conservation or protein homology.

2.4.1 Nucleotide Homology

If a well-annotated genome from a closely related species is available, it is possible to essentially utilize the gene models from that organism and the functional annotations to assist in the annotation of most regions of the new genome. This process is essentially very similar to that of the mRNA mapping process discussed earlier. This automatic annotation allows a focus on the traits that make the new organism interesting or unique (there must be something different in the new genome, otherwise why is it being sequenced?). Beyond point mutations that can be captured through mRNA-style mapping, the differences tend to include chromosomal rearrangements, gene duplications or deletions, inversions, and insertions of unique genes or stretches of genomic DNA. All of these variations from the already characterized genome require the calculation and visualization of large-scale genomic similarity (synteny).

Some synteny-based gene structure predictors such as NGENE (Arumugam et al., 2006) do not rely on preexisting accurate gene models in related species. Instead, they leverage the idea that most nucleotide substitutions occur in the third codon position, because such substitutions tend to not affect the amino-acid translation. In this way, gene structure alignment can be distinguished from overall genomic synteny because of the relatively high selective pressure on exons.

The ratio of nonsynonymous to synonymous codon substitution rates for nucleotides (dN/dS) has also been used for the detection of pseudogenes. Pseudogenes are nonfunctional remnants of old genes in the genome that occur regularly in higher eukaryotes, and therefore their detection is important for proper gene annotation. Pseudogene identification is also supported by the lack of proper stop codons, frameshifts within the gene model, and lack of transcripts, when gene expression information is available.

Using some of both the nucleotide sequence conservation and protein homology methods of cross-species comparison, respectively, is the Pseudogene Inference from Loss of Constraint (PSILC) software (Coin and Durbin, 2004). When the pseudogene contains a characterized Pfam (http://pfam.sanger.ac.uk) domain, PSILC calculates the probability that the nucleotide and amino acid substitutions are neutral with respect to the domain model. This is a more complex measure than just synonymous versus nonsynonymous.

2.4.2 Protein Homology

Pure protein-homology-based methods are very popular for gene prediction if only because of the massive wealth of protein data available in the public domain. Even when no closely related species has a full genome available, one can expect the majority of genes in a new organism to have some homology to known genes. These may be from closely related species in the case of parallel evolution or distantly related organisms in the case of horizontal gene transfer (also known as lateral gene transfer or LGT). For prokaryotic genomes, this homology-based gene prediction is done using a standard sequence alignment algorithm such as BLASTP (Altschul et al., 1997) to identify the core of the protein, followed by a search for an in-frame stop codon. The start codon choice can be supported by the same technique as *ab initio* prediction: the absence of rare codons, as well as the presence of –35 and –10 signals.

In Eukarya, a double-affine search is necessary. Parameterization of a double-affine protein-versus-genomic search is more subtle an art than mRNA-to-genomic alignment. This is because the protein alignment will inevitably have both small gaps due to protein evolution, and larger gaps representing introns. The number and size of these will depend on the evolutionary distance of the homolog and the characteristic intron size for the organism. Smart aligners will attempt to introduce long (not so penalized) gaps at acceptor and donor sites in the genomic sequence. One aligner that takes this a step further is GeneWise (Birney et al., 2004). GeneWise combines a pairwise matching algorithm with a Markov model not unlike *ab initio* predictors in order to help correctly place the introns into the alignment. This is especially helpful when the protein homology is weak, creating ambiguous choices for long gap insertion.

Some aligners, including GeneWise, will provide the option of either global or local alignment. If the only source for determining gene structure will be the protein homology, a global alignment ensures that the whole protein is being correctly modeled. If there are other sources for exon structure (e.g., short reads), a local alignment can assist in transferring over protein domain information rather than just the stricter gene ortholog information from the public databases. Protein alignment tools, which compensate for frameshifts due to genomic assembly errors, will also perform better when elucidating gene structures.

Another consideration is the similarity matrix used for protein-level matches. By default, most search methods use the BLOSUM62 matrix (Henikoff and Henikoff, 1992), which is lenient in allowing substitutions between various amino acids to be considered a "match." If the database contains closely related species, better alignments and statistics can be generated using the BLOSUM80 or the BLOSUM90 matrices. Higher numbers are stricter on amino acid substitutions, as explained further in Chapter 7.

Another caveat about protein homology searches is the quality of the protein database being used. The explosion in the number of species sequenced using next-generation DNA sequencing technology has inevitably led to poorer quality gene models being submitted to and accepted by databases and especially public data repositories such as GenBank (the submitter is after all responsible for the quality of the database submission), since less time or even no time is spent to manually examine the evidence for each gene model predicted by automated methods within the sequence. While this caveat is well known for functional genome annotation (discussed later in this chapter), the number of chimeric eukaryotic genes due to false assembly is less known, where two or more neighboring genes are fused into a single gene model in the annotation. This problem can be especially caused by the misconfiguration of genome assembly pipelines. Another reason might be the weak statistical differentiation between splice sites and start–stop patterns in many genomic regions. To counteract the risk of carrying over the same chimeric gene structure into a related genome, it is important to consider the entire set of significant protein matches rather than just the top one presented in the output of the database search tools.

2.4.3 Domain Homology

Related to the overall field of protein homology searches is a more subtle technique known as domain search. The results of these searches can provide better statistical matches than protein homology searches, when the gene in question is only distantly related to known genes. In a domain search, the database that it is searched against contains a set of hidden Markov models (HMMs) such as Pfam or the contents of the entire InterPro database (Hunter et al., 2009). In these models, a protein family is represented by a matrix, with separate amino acid substitution rates for each position in the domain. This is in contrast to pairwise protein homology searches using BLOSUM matrices, which are based on the assumption that the substitution rate is the same across the whole protein. The advantage of position-specific scoring is that it more accurately reflects the evolutionary history of a particular protein family, and hence match statistics will be better than a pairwise search against each individual member of the family. Of course, a double affine search is required when searching through eukaryotic sequences.

Two main caveats regarding domain searches are partial matching and spurious hits. Partial matching means that the whole structure of a gene is unlikely to be elucidated using a domain search in eukaryotes, because most protein domains only cover a portion of the total gene length. For example, the ATP-binding cassette (ABC) protein domain match from Pfam typically covers only 60% of a gene, making it likely that the alignment misses exons, which exclusively cover the remaining 40% of the sequence. As such, domain matches supplement, rather than replace, protein level searches. Spurious matches can also occur if small portions of the genome match highly conserved regions of a protein domain, but overall similarity is low. It is therefore advisable to judge domain hits not just by using the E-value (expected value), but also the percentage coverage of the domain length.

2.5 COMBINING GENE PREDICTIONS

Another class of gene prediction programs is those that combine evidence from multiple prediction tools into one consensus result. These are mostly eukaryotic prediction tools, which can be subdivided into two main categories: explicit scoring and automatic scoring. For explicit scoring tools, the user provides their preferences in order to weigh/rank evidence as input, coming from multiple sources, such as manual curation, *ab initio* prediction tools, promoter searches, and information such as mapped ESTs. Examples include AUGUSTUS (Stanke et al., 2006), JIGSAW (Allen and Salzberg, 2005), EuGene (Foissac et al., 2008), and geneid (Blanco and Abril, 2009). These programs generally require a training phase, during which a voting system, decision tree, or weight matrix is created. The advantage of this method is that the scoring system is relatively transparent and can be tweaked by the end user. On the other hand, automatic scoring systems generally require little to no training, relying on machine learning techniques to optimize weighting via dynamic programming or support vector machines and the like. The trade-off in using these tools is that the scoring systems tend to be relatively opaque to the end user, making the tweaking of the results, if not as expected, more difficult. Examples of such programs include mGene (Schweikert et al., 2009), GeneMark (Ter-Hovhannisyan et al., 2008), Evigan (Liu et al., 2008), and fgenesh (www.softberry.com).

In general, consensus prediction approaches, which employ multiple tools, produce more accurate gene models than any individual gene prediction program used alone. The main measures for accuracy cited in the literature are Sn (sensitivity, or 1 – False negative rate) and Sp (specificity, or 1 – False positive rate), often averaged out to give an overall measure of accuracy. Tables comparing performance of prediction tools can be confusing, as these measures can be calculated (in descending order of accuracy) at the nucleotide, exon, gene, and transcript levels. Although almost every tool can claim excellent performance in at least one of these categories, none is yet accurate enough to be taken as-is for complete gene structure prediction. Manual curation of gene models is still necessary to obtain optimal results, as it is quite difficult to impossible to capture the underlying biological knowledge that an expert can provide automatically with any of the approaches described. A prediction combination case study is explored in Chapter 11.

2.6 SPLICE VARIANTS

For many eukaryotic genes, there is in fact not just one correct gene model structure, but multiple versions are true, at least due to splice variants. The general categories of splicing events include (in descending order of frequency) skipped exons, alternative donors and acceptors, retained introns, and novel introns. Many gene prediction programs can be utilized to generate models of alternative transcripts, at least some of which will correspond to actual splice variants. The validation of alternate models, like that of the primary models, involves the mapping of mRNA data to the genome. Traditionally this was done using EST sequences elucidated by Sanger sequencing. To improve the depth of transcript coverage to find rare splice events, ESTs from related species could be mapped to the target genome, when genes were highly conserved. This resolution of variants can now be greatly enhanced by the spliced mapping of short next-generation sequencing mRNA data to the genome.

Skipped exons fall into two general categories: inner and outer. Outer skipped exons lead to proteins that are truncated at the N- or C-terminus, or sometimes the protein translation is unchanged and only the UTRs are affected. Figure 2.2 illustrates how two prediction programs disagree on the gene structure, but both are likely correct. GeneMark captures the form of the gene that skips the last exon (the stop codon is just inside what is normally the intron). AUGUSTUS identifies the more common form that includes the last exon.

Alternative splice donors and acceptors yield shorter or longer versions of a protein, and sometimes dramatically so, because of changes in the reading frame caused by the different splice sites. The same holds true for skipped inner exons, if their length is not a multiple of three. Alternate acceptors and donors are caused by the presence of multiple valid intron structures, which the spliceosome can excise. While splice repressor proteins are known, the complete mechanisms behind the preferential splicing of one form or another under various environmental conditions are not completely understood. Sometimes the distinction between alternatively spliced exons and skipped exons can be difficult. In Figure 2.3 there are likely two forms of the gene. One consists of exon x with donor site A, followed by exon y, but exon y starts at splice site C. The other form consists of x, followed by y, which spans splice sites B to D, then exon z starting at site E.

FIGURE 2.2   GBrowse view of spliced reads mapped to a genome. AUGUSTUS predicts an extra 3′ exon, supported by reads, but GeneMark also has some reads supporting an early gene termination in the intron (skipped last exon).

The *ab initio* predictors picked up strong acceptor (C) and donor (D) sites, but failed to determine that they belonged to different splice variants. Should the longer form of *z* be considered a splice variant with a skipped exon *y*, or is it a retained intron? The distinction is just a question of labeling.

Retained introns occur when the splicing mechanism fails to remove the intronic sequence by design. These introns are generally of a length that is a multiple of three, so that the reading frame of subsequent exons is not disturbed by the translation of the retained intron. Figure 2.4 displays a case where both variants are about equally expressed, but only one is predicted by the *ab initio* modelers AUGUSTUS and GeneMark.

The distinction between novel and retained introns is mostly one of perspective. If the most abundant mRNA includes the intron, other versions are retained. If it is more common to see the sequence as part of the mRNA, its skipping qualifies it as a novel intron.

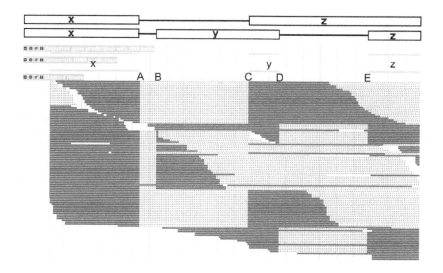

FIGURE 2.3 Alternate exon structures of a gene. Two likely splice forms of the gene exist (top). GBrowse view of spliced reads mapped to a genome are shown (bottom), with *ab initio* predictions above that erroneously capturing (C–D), the acceptor and donor of two alternate splicing forms.

In some cases, alternate splice forms of genes create so-called nonsense transcripts, which are never translated into protein. The ultimate validation of the protein expression of splice variants can be achieved through green fluorescent protein (GFP) fusions, mass spectrometry, and various other means. The fact that some variants are only expressed under very specific conditions can make this confirmation step all the more difficult.

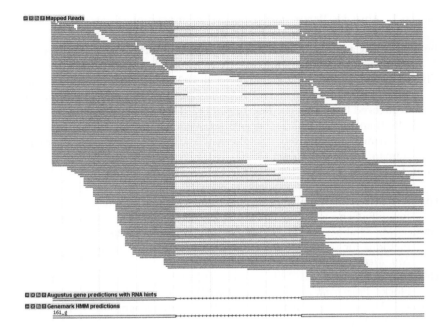

FIGURE 2.4 GBrowse view of spliced mapped reads and gene predictions. Clearly, the intron is retained in some versions of the transcript, as witnessed by the numerous unspliced reads in the predicted intron from AUGUSTUS and GeneMark.

REFERENCES

1000 Genomes Project Consortium. 2010. A map of human genome variation from population-scale sequencing. *Nature* 467:1061–1073.

Allen, J.E., Salzberg, S.L. 2005. JIGSAW: Integration of multiple sources of evidence for gene prediction. *Bioinformatics,* 21(18):3596–3603.

Altschul, S.F., Madden, T.L., Schäffer, A., et al. 1997. Gapped BLAST and PSI-BLAST: A new generation of protein database search programs. *Nucleic Acids Res.* 25(17):3389–3402.

An, J.J., Gharami, K., Liao, G., et al. 2008. Distinct role of long 3′ UTR BDNF mRNA in spine morphology and synaptic plasticity in hippocampal neurons. *Cell* 134(1):175–187.

Arumugam, M., Wei, C., Brown, R., Brent, M.R. 2006. Pairagon+N-SCAN_EST: A model-based gene annotation pipeline. *Genome Biol.* 7(Suppl. 1):S5.

Beaudoing, E., Freier, S., Wyatt, J.R., Claverie, J.M., Gautheret, D. 2000. Patterns of variant polyadenylation signal usage in human genes. *Genome Res.* 10:1001–1010.

Birney, E., Clamp, M., Durbin, R. 2004. GeneWise and genome wise. *Genome Res.* 14(5):988–995.

Blanco, E., Abril, J.F. 2009. Computational gene annotation in new genome assemblies using GeneID. *Methods Mol. Biol.* 537:243–261.

Butler, J., MacCallum, I., Kleber, M., et al. 2008. ALLPATHS: *De novo* assembly of whole-genome shotgun microreads. *Genome Res.* 18(5):810–820.

Chevreux, B., Pfisterer, T., Drescher, B., et al. 2004. Using the miraEST assembler for reliable and automated mRNA transcript assembly and SNP detection in sequenced ESTs. *Genome Res.* 14(6):1147–1159.

Clark, F., Thanaraj, T.A. 2002. Categorization and characterization of transcript-confirmed constitutively and alternatively spliced introns and exons from human. *Hum. Mol. Genet.* 11:451–464.

Coin, L., Durbin, R. 2004. Improved techniques for the identification of pseudogenes. *Bioinformatics* 20(Suppl. 1):i94–i100.

Foissac, S., Gouzy, J., Rombauts, S., et al. 2008. Genome annotation in plants and fungi: EuGene as a model platform. *Current Bioinformatics* 3:87–97.

Garber, M., Grabherr, M.G., Guttman, M., Trapnell, C. 2011. Computational methods for transcriptome annotation and quantification using RNA-seq. *Nat. Methods* 8:469–477.

Henikoff, S., Henikoff, J.G. 1992. Amino acid substitution matrices from protein blocks. *Proc. Natl. Acad. Sci. USA* 89(22):10915–10919.

Hertel, K.J. 2008. Combinatorial control of exon recognition. *J. Biol. Chem.* 283:1211–1215.

Homer, N., Merriman, B., Nelson, S.F. 2009. BFAST: An alignment tool for large scale genome resequencing. *PLoS One* 4(11):e7767.

Huang, X., Madan, A. 1999. CAP3: A DNA sequence assembly program. *Genome Res.* 9:868–877.

Hunter, S., Apweiler, R., Attwood, T.K., et al. 2009. InterPro: The integrative protein signature database. *Nucleic Acids Res.* 37(Database issue):D211–D215.

Iseli, C., Jongeneel, C.V., Bucher, P. 1999. ESTScan: A program for detecting, evaluating, and reconstructing potential coding regions in EST sequences. *Proc. Int. Conf. Intell. Syst. Mol. Biol.* 138–148.

Jacobs, G.H., Chen, A., Stevens, S.G., et al. 2009. Transterm: A database to aid the analysis of regulatory sequences in mRNAs. *Nucleic Acids Res.* 37(Database issue):D72–D76.

Kent, W.J. 2002. BLAT—The BLAST-like alignment tool. *Genome Res.* 12(4):656–664.

Kozak, M. 1986. Point mutations define a sequence flanking the AUG initiator codon that modulates translation by eukaryotic ribosomes. *Cell* 44(2):283–292.

Langmead, B., Trapnell, C., Pop, M., Salzberg, S.L. 2009. Ultrafast and memory-efficient alignment of short DNA sequences to the human genome. *Genome Biol.* 10(3):R25.

Li, H., Durbin, R. 2010. Fast and accurate long-read alignment with Burrows-Wheeler transform. *Bioinformatics* 26(5):589–595.

Liu, Q., Mackey, A.J., Roos, D.S., Pereira, F.C. 2008. Evigan: A hidden variable model for integrating gene evidence for eukaryotic gene prediction. *Bioinformatics* 24(5):597–605.

Oliver, J.L., Carpena, P., Hackenberg, M., Bernaola-Galván, P. 2004. IsoFinder: Computational prediction of isochores in genome sequences. *Nucleic Acids Res.* 32:W287–92.

Pribnow, D. 1975. Nucleotide sequence of an RNA polymerase binding site at an early T7 promoter. *Proc. Natl. Acad. Sci. USA* 72:784–788.

Qureshi, S.A., Jackson, S.P. 1998. Sequence-specific DNA binding by the *S. shibatae* TFIIB homolog, TFB, and its effect on promoter strength. *Mol. Cell* 1(3):389–400.

Rao, D.M., Moler, J.C., Ozden, M., Zhang, Y., Liang, C., Karro, J.E. 2010. PEACE: Parallel environment for assembly and clustering of gene expression. *Nucleic Acids Res.* 38(Web Server issue):W737–W742.

Robertson, G., Schein, J., Chiu, R., et al. 2010. *De novo* assembly and analysis of RNA-seq data. *Nat. Methods* 7(11):909–912.

Rosenberg, M., Court, D. 1979. Regulatory sequences involved in the promotion and termination of RNA transcription. *Annu. Rev. Genet.* 13:319–353.

Schatz, M.C. 2009. CloudBurst: Highly sensitive read mapping with MapReduce. *Bioinformatics* 25(11):1363–1369.

Schurr, T., Nadir, E., Margalit, H. 1993. Identification and characterization of *E. coli* ribosomal binding sites by free energy computation. *Nucleic Acids Res.* 21:4019–4023.

Schweikert, G., Zien, A., Zeller, G., et al. 2009. mGene: Accurate SVM-based gene finding with an application to nematode genomes. *Genome Res.* 19(11):2133–2143.

Sharp, P.M., Li, W.H. 1987. The Codon Adaptation Index—A measure of directional synonymous codon usage bias, and its potential applications. *Nucleic Acids Res.* 15:1281–1295.

Shine, J., Dalgarno, L. 1975. Determinant of cistron specificity in bacterial ribosomes. *Nature* 254(5495): 34–38.

Stanke, M., Keller, O., Gunduz, I., Hayes, A., Waack, S., Morgenstern, B. 2006. AUGUSTUS: *Ab initio* prediction of alternative transcripts. *Nucleic Acids Res.* 34(Web Server issue):W435–W439.

Ter-Hovhannisyan, V., Lomsadze, A., Chernoff, Y.O., Borodovsky, M. 2008. Gene prediction in novel fungal genomes using an *ab initio* algorithm with unsupervised training. *Genome Res.* 18(12):1979–1990.

Trapnell, C., Pachter, L., Salzberg, S.L. 2009. TopHat: Discovering splice junctions with RNA-seq. *Bioinformatics* 25(9):1105–1111.

Unniraman, S., Prakash, R., Nagaraja, V. 2002. Conserved economics of transcription termination in eubacteria. *Nucleic Acids Res.* 30(3):675–684.

Zerbino, D.R., Birney, E. 2008. Velvet: Algorithms for *de novo* short read assembly using de Bruijn graphs. *Genome Res.* 18(5):821–829.

CHAPTER 3

Between the Genes

3.1 INTRODUCTION

For much of the history of genome research, most of the sequence that was not protein coding in the human and other genomes was labeled "junk DNA." Besides the obvious exceptions of structural RNA genes, such as tRNA coding regions, as well as transcription factor binding sites, no function could be assigned to these vast stretches of genomic information. Not only is much of this extra sequence untranslatable into protein, it is also highly repetitive, lowering its information potential even more. Nevertheless, as the human genome was elucidated, it became clear that higher animals did not have many more genes than phylogenetically lower organisms, such as plants. The discovery of new classes of RNAs in the regions formerly labelled as "junk DNA" helped unravel new gene regulatory mechanisms. Today, we know that the genomic complexity in higher organisms is not based solely on gene structure, but on the intricate interactions between transcription factors, splicing factors, regulatory RNAs, DNA methylation, and chromatin rearrangements, as well as potentially other mechanisms, which have not yet been discovered. As such, genome annotation cannot be considered comprehensive without a thorough characterization of the genomic regions "between the genes."

3.2 TRANSCRIPTION FACTORS

Transcription factors (TFs) are usually regulatory proteins that bind to the genomic DNA alone or in conjunction with other proteins in order to enhance or suppress the rate of downstream gene transcription. These enhancement or repression activities can also work in cascades to greatly amplify or eliminate transcription of specific gene sets, or to strictly maintain transcriptional stasis, respectively. The binding of transcription factors to genomic DNA is complex, dependent on the three-dimensional structure of the protein complex, as well as the steric and biochemical characteristics of the genomic transcription factor binding site (TFBS, also known as response element). The electrostatic and van der Waals forces that dominate these interactions tend to allow TF recognition of several DNA sequence variants, and the general scheme of recognized sequences is known as a TFBS motif. More importantly, the presence of a TFBS motif by itself is insufficient to conclude that a gene is regulated by a specific transcription factor. To reduce the number of false positive TFBS calls, the scoring of the match, the location of the putative TFBS, the sequence neighborhood, and TFBS conservation across species should all be considered.

3.2.1 Transcription Factor Binding Site (TFBS) Motifs

Because any given TF can bind to a number of similar DNA sequences, recognition motifs are generally represented computationally as a position-specific weighted scoring matrix. Given a set of experimentally confirmed TFBSs for a particular TF, these sequences are aligned to build a matrix weighted toward highly conserved nucleotide positions. As an example, Figure 3.1 shows the alignment of known binding sites for the CREB1 transcription factor (Bartsch et al., 1998). Underneath this is the position-specific weighting of the matrix, based on column totals from the alignment. At the bottom is a common visual representation of the weighted matrix, known as a WebLogo (Crooks et al., 2004). Intuitively, the height of the letters is proportional to their importance in achieving a motif match in a scanned genomic sequence.

The validated TFBS sequences, their alignment, and the selection of columns to retain have an impact on the final matrix. Figure 3.1 represents these choices for the CREB1 TF in the JASPAR database (Sandelin and Wasserman, 2004). JASPAR is the most popular, fully public database of TF-weighted scoring matrices for eukaryotes. Many TFBSs for JASPAR are drawn from the Eukaryotic Promoter Database (Schmid et al., 2006), which documents experimentally validated elements. The PAZAR Web site (Portales-Casamar et al., 2007) also acts as an aggregate portal for public TF information. The TransFac database (Matys et al., 2006) has both public (not updated since 2005) and commercial versions, with the latter including many more matrices, which are mostly based on literature mining. MatBase (Genomatix Software GmbH, Munich, Germany) is a commercial database that also contains many TF scoring matrices, which

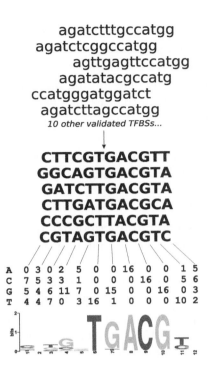

agatctttgccatgg
agatctcggccatgg
    agttgagttccatgg
    agatatacgccatg
ccatgggatggatct
agatcttagccatgg
10 other validated TFBSs...

CTTCGTGACGTT
GGCAGTGACGTA
GATCTTGACGTA
CTTGATGACGCA
CCCGCTTACGTA
CGTAGTGACGTC

A	0	3	0	2	5	0	0	16	0	0	1	5
C	7	5	3	3	1	0	0	0	16	0	5	6
G	5	4	6	11	7	0	15	0	0	16	0	3
T	4	4	7	0	3	16	1	0	0	0	10	2

FIGURE 3.1 (Top) Aligned sequences recognized by the CREB1 transcription factor. (Middle) Weighted matrix of sequence conservation. (Bottom) A sequence logo representing the relative conservation of bases by height.

are also based on in-depth literature mining. The greater depth of the commercial data sets should not be underestimated, if TF annotation is essential for a particular genome project. MatBase can be searched when subscribing to a license for the MatInspector software (Cartharius et al., 2005). Searching JASPAR can be accomplished using TOUCAN (Aerts et al., 2005) or similar software. TOUCAN has the advantage of also integrating several small, species-specific matrix libraries.

TF matrix matching can generate many false positives. A simple way to reduce the number of matches is to restrict the matrix libraries used to only those applicable. For example, most search tools allow the user to search only a subset of the matrices, for example, the plant TF matrices. Some statistics from TF matrix-to-genome matches can also help the annotator to exclude likely false positives. First, any search tool will generate a raw score for the match (M).

It will also assign a p-value matrix match, indicating the probability that M was found randomly. Lower p-values are more likely to be significant.

The p-value is based on some (perhaps sometimes incorrect) assumptions about the background frequency of nucleotides in the query. A more accurate p-value can be achieved by searching a random sample of sequences against the same matrix and determining the frequency of matches with as good a raw score as M. In the case of a genome annotation, the random sample would come from the set of all known 5′ untranslated regions (UTRs). Even with accurate match p-values, functionality cannot be assumed, simply because many other factors, besides the sequence motif recognition, may affect the TF binding.

3.2.2 TFBS Location

Transcription factors effect transcriptional regulation based on their interaction with the transcription initiation machinery of the cell. This means that TFBSs are generally located either near the transcriptional start site in the genome (proximal TFBSs), or they are brought near the start site through the three-dimensional folding of the genomic DNA (distal TFBSs). In most proximal TFBSs a number of "location-based clues" are available, which can be used to aid in their correct identification. First, the transcriptional start site (TSS) may be known through the mapping of mRNA data to the genome (see Chapter 2). This is especially true where deep short-read next-generation sequencing has been performed. Most proximal TFBSs can be found within 100 bases (upstream or downstream) of this site (Xie et al., 2005). In addition, in species with widespread DNA methylation, TFBSs tend to be rich in cytosines and guanines. Therefore, where the actual location of the TSS is unknown, proximity to guanine and cytosine (G+C)-rich stretches of about 200 bp (also known as CpG islands) can be a clue that a matrix match has found the true proximal TFBS. CpG islands are hotbeds for DNA methylation, another layer of transcriptional control (see Chapter 4).

The location of CpG islands is usually within 1000 bp of the start codon (Shimizu et al., 1997), therefore correct start codon prediction (see Chapter 2) can also be important to putative TFBS annotation. Consequently, genes with multiple real start codons (gene isoforms) may also have multiple sets of TFBS predictions.

3.2.3 TFBS Neighborhood

A transcription factor may act in concert with transcriptional cofactors to achieve specific activation or suppression of the transcription of certain genes. In this case, it can be expected to find particular combinations of

TFBSs in a relatively small 5′ region of the respective gene. Several software tools exist, which can be used to perform exactly this search, given known cofactor combinations (also known as cis-regulatory modules or CRMs). For example, Comet (Frith et al., 2002) uses the joint probability of user-selected TRANSFAC match co-occurrences to generate an e-value for the most probable TFBSs. A difficulty with this and similar programs is that the detection of motifs is highly dependent on the selection of input parameters, such as the expected gap between TFBSs, and factors such as sequence background characteristics. Cluster-Buster (Frith et al., 2003) provides a parameter optimization script that can help to reduce the burden for the end user trying to annotate TFBSs. RSAT (Thomas-Chollier et al., 2008) uses unique statistical methods such as adaptive background models and Markov-chain estimation of p-values. In most cases, the user must preselect a small list of TFBS matrices to scan for. By contrast, TOUCAN includes the ability to find overrepresented combinations of any TFBSs in a genome in order to discover TF cofactors *de novo* by running the built-in MotifScanner+Statistics tools, followed by the ModuleSearcher tool. CisModule (Zhou and Wong, 2004) can be used to detect modules without a reference TFBS data set, with the caveat that false positives are quite likely with this approach.

3.2.4 TFBS Conservation

Given the importance of transcription factors for the proper regulation of gene expression, it stands to reason that TFBSs should be conserved across closely related species. Although sequence variability in the recognition sites may complicate the task, the conservation rate is still above the general 5′ UTR conservation rate. EEL (Hallikas et al., 2006) is a popular CRM scanner that takes advantage of this rate differential, by aligning across species only the sequence fragments (in genomic order) that contain known TFBS motifs. oPOSSUM (Ho Sui et al., 2007) can also take advantage of cross-species conservation for certain precalculated species pairs, such as *Caenorhabditis elegans* and *Caenorhabditis briggsae,* or the human and mouse genomes.

The *de novo* detection of conserved TFBSs can be divided into two main categories: within-species mining and cross-species mining. Within-species mining relies on the fact that a transcription factor may be used to regulate more than one gene, and therefore similar TFBSs might be found upstream of multiple genes. The classic software for detecting recurring motifs is the MEME suite (Bailey et al., 2009). The basic technique of motif discovery requires the generation of a set of sequences, which are assumed to be coregulated by the same transcription factor. Weeder (Pavesi et al., 2004) more specifically targets TFBSs, and provides statistical "advice," based on automatically scanning through many possible interactions with different parameters. *De novo* discovery of TFBSs may be especially important in prokaryotic genome annotation or lower eukaryotes, since most existing databases are biased toward eukaryotic model organisms.

The computational intensity of these *de novo* algorithms makes most of them currently impractical for genome-wide TFBSs prediction, but other sources of data can aid in narrowing the search area. In particular, chromatin-immunoprecipitation sequencing (ChIP-Seq) data can be used as an additional input for several TFBS discovery tools. The Hybrid Motif Sampler (HMS) (Hu et al., 2010) can be used to search regions with high ChIP sequencing depth (i.e., lack of chromatin) and account for interdependencies of clustered motifs. ChIPMunk (Kulakovskiy et al., 2010) uses ChIP-Seq data, plus novel motif discovery heuristics, to dramatically cut the computational time but produces similar results to the performance of HMS. From a genome-wide prediction perspective, this compares favorably to gene-expression enhanced motif searches (Conlon et al., 2003), where the added predictive power is limited to the TFs correlated to the given experimental condition.

De novo sequence prediction using conservation between species is referred to as phylogenetic footprinting. ConSite (Sandelin et al., 2004) is one of the major tools in which cross-species TFBS filtering is implemented. Most tools, including ConSite, rely on the user providing an accurate mapping of orthologous genes as input, therefore accurate gene structure modeling (see Chapter 2) is a prerequisite for this type of analysis.

3.3 RNA

Ribosomal nucleic acid (RNA) was for many years considered DNA's poor cousin. Whereas genomic DNA held the blueprints to life, RNA was thought of as a simple messenger and structural component of the cell. Beyond these traditional roles, measurements of sequence-specific interactions between RNA–RNA and RNA–DNA have revealed highly complex genetic regulation mechanisms in eukaryotes and to a lesser extent in archaea. Although our knowledge of the partners in the RNA interaction is currently incomplete, the computational tools described next should remain largely applicable for the prediction of the specific RNA sequences involved. A relatively comprehensive database of known noncoding RNAs covering most of the categories is Rfam (Gardner et al., 2011).

3.3.1 Ribosomal RNA

The ribosomal RNA (rRNA) genes are structural constituents of the backbone of the ribosome. They may be encoded more than once in a genome from any branch of life. Typically, the rRNA sequences show a high G+C content (approximately 65%). The distribution and total number of RNA genes depend on the branch of life being examined. They are named according to their sedimentation properties (S or Svedberg units). In prokaryotes, the three rRNA (5S, 16S, and 23S) are usually organized in an operon (cotranscribed). Accurate identification of the 16S rRNA sequences is important to determine the ribosomal binding site discussed in gene structure prediction (see Chapter 2). In eukaryotes, the four rRNAs (5S, 5.8S, 18S, and 28S) are typically present in many copies per genome. The 5S location is transcribed by RNA polymerase III. It is located separate from the other RNA genes, which are cotranscribed by RNA polymerase I. The 5.8S, 18S, and 28S genes are separated by so-called internally transcribed spacers (ITS). ITS sequences are often used for phylogenetic fingerprinting of closely related species and are therefore of interest to researchers. Almost all eukaryotes also contain two mitochondrial rRNA genes. rRNA sequences do not necessarily reflect the final organization of the structural element in the ribosome, as the rRNA precursor molecules may undergo cleavage and various nucleotide modifications after transcription. RNAmmer (Lagesen et al., 2007) provides a sensitive hidden Markov model search against a prebuilt set of rRNA models. Alternatively, rRNA genes can be identified using nucleotide sequence homology to reference rRNA sequences from Rfam (Gardner et al., 2011).

3.3.2 Transfer RNA

A transfer RNA (tRNA) is an approximately 76-bp-long RNA molecule, responsible for concatenating an amino acid to the C terminus of a polypeptide chain in protein synthesis. Each DNA codon has its own tRNA, but "wobble" in the third codon base means that an organism may have less (and of course many need less) than the 61 ($4^3 - 3$, excluding 3 stop codons) tRNAs. In prokaryotes there is generally one copy of each tRNA, whereas hundreds of copies can exist in some higher eukaryotes. tRNAs have a canonical cross-like secondary structure as shown in Figure 3.2, therefore tRNA prediction algorithms are mostly based on finding sequence segments that may fold into this secondary structure.

Free energy calculation of folding is a common approach to predicting these structures but is complicated by possible noncanonical base pairing

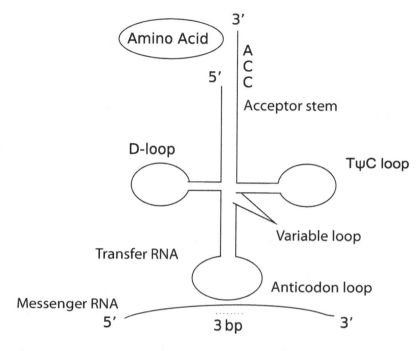

FIGURE 3.2 Canonical two-dimensional structure diagram for a transfer RNA, with a cross-like set of stems, loops, and bulges.

in the acceptor loop. Another sequence feature expected (in eukaryotes) is the RNA polymerase III binding site 8 near the start and just over halfway through the tRNA gene. Mature tRNAs are derived from longer pre-tRNA sequences, which are found in the genomic sequence. This modification can both aid and hinder the prediction process. On the plus side, the pre-tRNAs transcripts are terminated by a stretch of four or more thymidines. On the negative side, the pre-tRNAs may contain introns. Introns in bacteria are usually self-splicing ribozymes. In archaea, introns contain a bulge–helix–bulge (BHB) secondary structure that is recognized by a splicing endonuclease. This motif, which can be between 11 and 175 bp long, can occur almost anywhere in the pre-tRNA. Some tRNAs may contain two (or sometimes even three) introns. Eukaryotes may contain BHB motifs, but more often have an intron adjacent to the anticodon, recognized by a eukaryal splicing endonuclease. The BHB motif consensus is species specific and varies considerably, therefore many prediction programs have the taxonomic division of the input as a parameter. The classic genomic tRNA prediction tool is tRNAScan (Lowe and Eddy, 1997). ARAGORN

(Laslett and Canbäck, 2004) uses homology to existing tRNAs and heuristics to detect tRNAs and transfer–messenger RNAs (bacterial tRNA-like structures that contain an internal reading frame that can effectively truncate protein synthesis).

The discovery of so-called split tRNAs has required the creation of new tRNA searching algorithms for archaea. Split tRNAs consist of two halves, which are merged subsequently into the mature pre-tRNA. The two halves are positioned at physically separate locations in the genome and independently transcribed. Perfectly complementary G+C-rich sequences, flanking the halves (3′ of the first half and 5′ of the second half), bind to form a BHB-like motif that is subsequently excised like a typical archaeal tRNA intron. SPLITS (Sugahara et al., 2006) detects possible split and BHB-containing tRNAs, removes the intervening sequences, then performs a standard tRNAScan to make a final prediction.

More recently, so-called permuted tRNAs in algae and archaea have been discovered that have the first and second halves of the tRNA reversed with 7 to 74 intermediate bases. These pre-tRNAs likely form a circular molecule that is then cleaved near the anticodon to form a canonical mature tRNA. Split and permuted tRNAs can be detected using an improved version of tRNAScan (Chan et al., 2011).

Mitochondrial tRNAs often do not follow the canonical cross structure and are therefore missed by traditional tRNA scanners. The ARWEN package (Laslett and Canbäck, 2004) is specifically designed to detect these unusual gene structures. Although the canonical tRNA structure is relatively strict, the presence of numerous pseudo-tRNA genes in eukarya must also be taken into account when performing tRNA annotation. Is the anticodon already covered by a much higher-scoring prediction elsewhere in the genome? Is it a species that might have split or permuted tRNAs that were thus far not searched for?

3.3.3 Small Nucleolar RNA

Small nucleolar RNAs (snoRNAs) guide the chemical modification of other RNA sequences, primarily the ribosomal RNA. The RNA target of a snoRNA is based on the complement of an 11 to 20 bp recognition pattern in the snoRNA sequence. SnoRNAs fall into two major classes, depending on their folding pattern: H/ACA or C/D box. H/ACA snoRNAs can vary in length from 70 to 250 base pairs and are typified by up to three helix-internal loop–helix–apical loop units. Figure 3.3 shows the structure of one such unit. Because of the threading of the target RNA into this

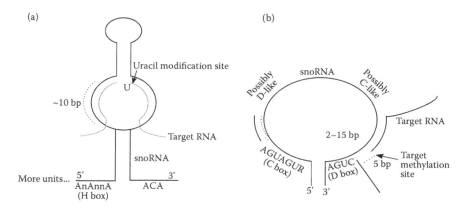

FIGURE 3.3 (a) Canonical two-dimensional structure diagram for an H/ACA snoRNA, with multiple units of the form helix-internal loop–helix–apical loop. (b) Two-dimensional structure diagram for a C/D snoRNA, with a small stem and a large internal loop.

structure, the recognition site is split into two parts, with the modification site (typically causing pseudouridylation) at 10 bases into the recognition site. The chemical modification is done by the snoRNA in a complex with four highly conserved proteins common to eukaryotes and archaea. In addition to the polyuridylation of rRNA, H/ACA complexes are also known to modify snRNAs and the telomerase ribonucleoprotein complex. Over 100 snoRNA targets are known in the human genome alone and additional snoRNAs without obvious RNA targets are known as orphans.

The search for H/ACA structures is rather computationally expensive, as whole genomes should be processed in segments to successfully perform this kind of prediction. Input to snoRNA prediction software is the genomic segment to be searched, the possible set of RNA target sequences, and in some cases the phylogeny of the input sequence since archaeal and eukaryotic snoRNAs may be modeled differently. SnoGPS (Schattner et al., 2004) uses a specialized target alignment algorithm and structure prediction heuristics to predict H/ACA snoRNAs. Fisher (Freyhult et al., 2008) uses a free-energy minimization approach to predict H/ACA-like structures. In either case, the small, split-target RNA recognition site can lead to the prediction of numerous false positives. The RNAsnoop software (Tafer et al., 2010) reduces false-positive predictions by filtering its initial predictions with a statistical method known as a support vector machine (SVM). The provided SVM is initially trained using confirmed C/D box snoRNA sequences. Evidence for target polyuridylation, or targeted snoRNA sequencing, can be used to confirm such predictions. Barring experimentation, sequence-based evidence to support prediction of snoRNAs can be obtained through the presence of an RNA polymerase II or III binding site upstream of the predicted motif, or more commonly, the placement of the location of the predicted motif inside the intron of an actively transcribed gene.

The second category of snoRNAs is the C/D box family. These snoRNAs form a complex with four proteins and normally cause methylation of target RNA molecules. The C box (consensus RUGAUGA) is located near the 5′ end of the snoRNA, while the D box (consensus CUGA) is located near the 3′ end (Kiss, 2002). The general secondary structure is a small stem with a large internal loop (see Figure 3.3). The contiguous nature of the target RNA recognition site makes it relatively easy to correctly identify C/D box snoRNAs, when compared to their H/ACA cousins. The original C/D box snoRNA prediction tool was snoscan (Lowe and Eddy, 1999), which is still widely used. It is based on a probabilistic sequence model. The PLEXY software (Kehr et al., 2011) rapidly predicts C/D box snoRNAs on a genomic scale, using a dynamic programming algorithm. The rate of false positives for this tool is very low.

Some tools conveniently combine H/ACA and C/D prediction systems into a single package, such as snoSeeker (Yang et al., 2006) and the RNAz Web server (http://rna.tbi.univie.ac.at/cgi-bin/RNAz.cgi). While most tools expect target RNA to be provided, the snowReport (Hertel et al., 2008) tool can provide both types of snoRNA predictions based on structure alone, with a reasonable rate of false positives using SVM filters as well.

In eukaryotes, besides snoRNAs, another family of RNAs is found in the nucleus whose members associate with proteins to form the machinery for mRNA splicing, the spliceosome. These small nuclear RNAs (snRNA), named for their ribonucleoprotein complex are called U1, U2, U4, U5, and U6. While there are variants within these designations, they are all approximately 150 bp in length and can be detected by DNA homology to Rfam (Gardner et al., 2011).

3.3.4 MicroRNA

MicroRNAs (also known as miRNAs) are RNAs about 22 bp long that form an imperfectly paired hairpin structure. They are found in the genomes of eukaryotes and affect protein production via a mechanism known as RNA interference (RNAi). In a complex with several proteins (most notably the Dicer RNA endonuclease), miRNA molecules guide the cleavage of mRNA at a location roughly complementary to the first 6 to 8 bp at the 5′ end of the miRNA. This cleavage either blocks the mRNA from being translated or promotes mRNA degradation. In terms of genome annotation, of primary interest to the researcher is of course the target gene suppressed by the miRNA. The complementarity of the 5′ miRNA end is often inexact, relying on DNA thermodynamics and steric availability to efficiently target mRNA (Silva et al., 2003). The target site is also commonly, but not exclusively, located in the 3′ UTR of the target mRNA. The complexity of the match between the miRNA and its target has led to the development of several bioinformatics approaches to their discovery. The first method relies on observing the conservation of the miRNA among several species. Tools that are based on this method scan intergenic regions for conservation (which are normally highly divergent) and filter for roughly palindromic subsequences (Grün et al., 2005). The identification of false-positive palindromic sequences, which are complementary to 3′ UTRs, can be limited by requiring adenosine base composition bias in the mRNA flanking sequence (Lewis et al., 2005). The authors of the PITA software (Kertesz et al., 2007) developed the flanking sequence filter further, using a two-step model where initially the 6 to 8 bp complementary seed must match. After the seeding, the criteria are the thermodynamic accessibility of the mRNA and the RNA stability of the miRNA–mRNA complex. This model adequately explains the different efficiencies of various miRNA-target combinations.

MiRNAs may constitute an early form of something similar to the innate immune system in vertebrates, which defends these organisms against double-stranded RNA viruses. Viral RNA is broken down into pieces of approximately 20 bp. These small interfering RNAs suppress viral gene transcription using the same mechanism as miRNAs. In bacteria and especially archaea, a system reminiscent of miRNA is used to defend against phages. It constitutes an acquired, rather than innate, and limited form of an immune response system. In these organisms, arrays of regulatory repetitive sequences called clustered regularly interspaced short palindromic repeats (CRISPR) elements can store fragments of phage DNA to which the cell has previously been exposed. When re-exposed to the phage, the cell splices out the fragment from the CRISPR array in the genome. The fragment associates with Cas proteins that degrade complementary phage DNA or mRNA. The identification of CRISPR elements can be achieved based on spaced repetitive elements and similarity to known phage repetitive elements. A database of CRISPR arrays in prokaryotic genomes (which have been published) is maintained by CRISPRdb (Grissa et al., 2007). In the case where it is necessary to discover CRISPR elements *de novo* in a new genome, PILER-CR (Edgar, 2007) provides good sensitivity and specificity.

3.3.5 Other Types of RNAs

Another class of RNAs found in animals is known to interact with the Piwi protein (also known as piRNAs). These molecules are typically 25 to 30 bp long and were first discovered to be active in germline cells, playing a role in the repression of retrotransposon activity. Tens of thousands of piRNA species have been found in insects, fish, and mammals. These short sequences map back to the genome in clusters of ~10 kb, which suggests that a small number of common precursor mRNAs are initially cleaved to form them. A database of known piRNAs is available (Sai Lakshmi and Agrawal, 2008), but piRNAs are not highly conserved across species (Ro et al., 2007), therefore the annotation of piRNAs in new species is primarily based on sequencing experiments. The precursors are typically found in repetitive intergenic sequences in the genome. PiRNA-like sequences are found throughout nongermline cells (Yan et al., 2011) and are now believed to play a role in epigenetic regulation, positive regulation of translation, and improved mRNA stability.

An mRNA may contain self-regulatory elements. Although these are most commonly found in bacteria, they are also present to some degree in archaea and eukaryotes. Cis-acting elements, which can be computationally identified, include selenocysteine signals (Chaudhuri and Yeates, 2005) and programmatic translation frameshifts −1 (Theis et al., 2008) as well as +1 (Liao et al., 2009). In the 5′ and 3′ UTRs of genes, the RNA may be able to bind small molecules and regulate the mRNA expression. These are known as riboswitches. For example, the SAM-I riboswitch controls protein levels of genes involved in methionine synthesis based on the concentration of S-adenosylmethionine (SAM) in the cell. These patterns can be found using a Riboswitch Finder Web server or the equivalent standalone software (Bengert and Dandekar, 2004).

Many long noncoding RNAs (lncRNAs) regulate genes in eukaryotes. These lncRNA genes are transcribed but not translated into proteins, hence they should really be named long non-protein-coding RNAs to be precise. Antisense RNAs are widespread, with most of them down-regulating gene activity. Antisense RNA can bind to complementary mRNA, which results in the formation of double-stranded RNA that is actively degraded by the cell. Antisense RNAs can also bind DNA. A well-known antisense RNA is Xist, which binds to and inactivates one copy of the mammalian X chromosome in females. A comprehensive database of known long noncoding RNAs is available on the Web (Amaral et al., 2011), with a focus on mammals.

Last, guide RNAs (gRNAs) mediate the posttranscriptional editing of mRNAs in certain protists, which contain special mitochondrial-packed DNA (kinetoplastids). A model for the transcript-specificity of tens of thousands of these RNAs has been proposed (Reifur et al., 2010).

3.4 PSEUDOGENES

During the course of genome evolution, some genes were duplicated. Subsequently some of the copies encountered mutations, which often rendered them nonfunctional. The mutation events may lead over time to significant truncation of the protein product through the introduction of in-frame stop codons. Alternatively, mutations in the regulatory regions that control gene transcription or splicing can render a protein effectively untranslatable. In either case, the result is referred to as a pseudogene. Pseudogenes are a pervasive feature in many prokaryotes, appearing and disappearing relatively quickly through genome evolution (based on the low number of pseudogenes that are shared across taxa). The exact reasons for prokaryotic pseudogenes to be retained are not known, but a higher number are found in recently evolved pathogens (Lerat and Ochman, 2005).

In eukaryotes, pseudogenes can exhibit influence on gene expression, and play a role in the generation of genetic (e.g., immunological) diversity (Balakirev and Ayala, 2003). Functional roles for pseudogenes are implied by the fact that sequence divergence from the functional genes is lower than randomly expected, and that single nucleotide polymorphisms are biased toward being synonymous versus nonsynonymous. Pseudogenes may no longer translate into protein, but may still affect their parent protein in some way. For example, mice without the makorin1-p1 pseudogene exhibit severe developmental abnormalities (Yano et al., 2004), because makorin1-p1 is expressed and confers stability to the protein-coding mRNA for makorin1. This case gives credence to the idea that annotation of eukaryotic pseudogenes is important, because in some cases sequence variants of pseudogenes can cause phenotypic or other changes.

Detection of inactivated, but otherwise intact, genes requires a thorough knowledge and annotation of the transcribed gene products and the regulatory regions of a gene. For accuracy, this is primarily a manual task that requires judging evidence such as the disappearance of CpG islands or known TFBSs for a gene. Systematic detection of all other types of pseudogenes is done using pairwise homology search of annotated genes against the genome. In order to properly identify pseudogenes, it is important to understand how they come about and consequently the types of structures they may have relative to their source gene. Pseudogenes may fundamentally be duplicates of genomic segments, or processed mRNAs, which were integrated into the genome via a reverse transcriptase. In the case of genomic segment duplicates, the original exon segmentation is often largely conserved. Searching for these pseudogenes requires either searching known protein-coding DNA sequences against the genome with an intron-tolerant alignment program or searching known exons against the genome with a regular BLAST-type search. Some paralogs are turned into pseudogenes by stop codons, which were introduced by point mutations. The older these pseudogenes, the more likely it is that pseudogene exons will have been lost. In other instances, the gene duplication is only partial. In either case, the search algorithm should only expect conservation across part of the total length of the source gene.

Processed pseudogenes on the other hand, which were derived from mRNA, have no introns. Initially, truncation of the gene in this case is more likely at the 5′ end of the gene, because reverse transcription starts at the 3′ end of the mRNA. The mRNA's poly-A tail and polyadenylation signal are in fact sometimes retained in these pseudogenes as well. As processed pseudogenes age, 3′ sections are also potentially lost through genetic drift or recombination. The search for processed pseudogenes requires only a BLAST-type search of known genes against the genomic DNA. The set of known genes to search against depends on the level of analysis desired and available compute resources. The set of all annotated full-length genes is a clear starting point for a systematic revelation of gene duplicates (with the variations described earlier) and processed pseudogenes. It is also possible that no intact source gene remains in the genome, in which case the inclusion of genes from related species may have to be used as the alignment template. The fact that protein-level conservation is often higher than DNA conservation in pseudogenes makes protein-level alignment to the genomic sequence desirable. The percentage identity cutoff for the correct annotation of a pseudogene is also an issue; as the percentage drops below about 70%, only the source gene family, not the particular source gene, may be evident without thorough phylogenetic analysis. Also, if only a small portion of a source gene remains, it is unlikely to have any remaining biological function. Setting a minimum threshold for the percentage of source gene length limits the annotation task, but in some species it can still be a daunting task; for example, in humans there are over 2000 ribosomal protein pseudogenes alone, with greater than 70% of the source gene length (Zhang et al., 2002).

An additional complication for any type of pseudogene search is the potential for retrotransposon insertion in the middle of the pseudogene. This insertion may be the original cause of the gene disablement or a subsequent event. In either case, the result is two pseudogene fragments.

3.4.1 Transposable Elements

Transposable elements are relatively short genetic sequences, which are copied or moved around a genome. The two major groupings, based on whether the copying mechanism requires intermediate RNA or not, are retrotransposons and DNA transposons, respectively. DNA transposon sequences that have been characterized are typically labeled as "IS" (insertion sequence) elements. Structurally, most IS elements contain a transposase gene, and flanking inverted repeat sequences. These sequences are found throughout the tree of life. In eukaryotes, many transposase genes are inactive, due to nonsense mutations. Inactivation mutation can be reversed in some instances to create "domesticated" transposase genes used by the host genome (Volff, 2006). Retrotransposon activation can be linked to disruption of normal genomic function and hence is often linked to disease (Kazazian, 1998). On the other hand, IS elements can be quite active and constitute an important mechanism for prokaryotic evolution (Blot, 1994). In eukarya, DNA-transposons are linked to large-scale structural changes and epigenetic modifications (Feschotte and Pritham, 2007).

3.4.2 DNA Transposons

Identification of DNA transposons can be achieved by homology search against known transposase genes. Where the transposase has not been active for some time, protein similarity may be weak. In this case, better sensitivity can be achieved using PSI-BLAST, an iterative form of the BLAST protein alignment algorithm. The TransposonPSI software (http://transposonpsi.sourceforge.net) identifies many types of transposable elements in this way. Identification of degraded DNA transposons can also be done using hidden Markov models, though care should be taken by the training of these models using special statistical techniques. This helps to ensure that the model is not too tied to the training set given (Edlefsen and Liu, 2010). While most DNA transposons are so-called cut-and-paste elements based on their transposase-mediated relocation mechanism during cell division, two other types of DNA transposons also exist: helitrons and polintons (Jurka et al., 2007). Helitrons are structurally related to bacterial IS97 with a helicase gene and are therefore believed to transpose via the same mechanism, the so-called rolling circle process. Helitrons are found in various eukaryotes but are particularly known to be very active in plants such as maize, where helitron-derived sequences compose approximately 2% of the genome (Feschotte and Pritham, 2009). Helitrons have a preference for transposing into other helitron copies, near but not within genes. Homology search and structural annotation can help identify these transposable elements.

The third type of DNA transposon, polintons, is found throughout eukaryotes (Kapitonov and Jurka, 2006). Polintons contain all of the genes required for their own transposition and are therefore also called self-synthesizing transposons. The set of genes includes a protein-primed DNA polymerase B, a retroviral integrase, a cysteine protease, and an ATPase. This sequence is likely derived from an ancient plasmid, but the unique combination of cis-genes makes it relatively straightforward to detect by using protein-homology methods. Additionally, polinton flanking inverted repeats are typically several hundred nucleotides long, capped 5′ by AG and 3′ by TC.

3.4.3 Retrotransposons

Many transposable elements are mobile based on transcription to RNA, followed by reverse transcription back to DNA and insertion back into the genome; these are called retrotransposons (Jurka et al., 2007). Because this action is copy and paste, rather than DNA transposon cut and paste, these elements tend to proliferate and enlarge the genome. The main distinction between various retrotransposons is whether they are of retroviral origin.

Retrotransposons of retroviral origin are typified by the presence of non-protein-coding long terminal repeats (LTRs) flanking protein-coding genes. In fully active retroviruses, the 5′ flank consists of four regions: R, U5, PBS, and L. The R region is a short repeat sequence found at both ends of the virus, used as a checkpoint during polymerization. The U5 sequence that follows is unique. This is followed by a primer-binding site (PBS), which is complementary to the 3′ end of a tRNA primer in the viral host. Sometimes the corresponding tRNA name is used in the nomenclature for naming the virus. Finally, the 5′ leader region (L) contains a signal important for the packaging of the genomic RNA. The viral 3′ flank includes a polypurine tract (PPT; i.e., polyA/G), which primes DNA synthesis during reverse transcription. This is followed by the proviral transcriptional primer U3 and finally R.

The core genes of any retrovirus are referred to as *gag*, *pol*, and *env* (Jurka et al., 2007), and are always found in this order. Gag proteins are cleaved, typically by a viral protease, into "mature" peptides forming essential virus particle structures: the matrix, the capsid, and the nucleocapsid. Pol proteins encode nucleic acid processing enzymes covering important roles in the life cycle of the virus, such as reverse transcription (polymerase), genome integration (integrase), and RNase H activity. The Env protein encodes the viral envelope protein.

Integrated retroviruses that can still be transcribed are referred to as endogenous retroviruses (ERVs), with the intergrated DNA referred to as a provirus (Gifford and Tristem, 2003). Retrotransposons near genes can affect their transcription. ERVs expression has also been linked to a host of chronic diseases, therefore their annotation is of more than academic interest. Depending on how long the retrovirus has been integrated in the genome, many parts of the original retroviral structure may be missing or mutated too much to be easily detectable using a BLAST-based method. LTR_STRUC (McCarthy and McDonald, 2003) can be used to detect retrovirus-like elements, which are flanked by long terminal repeats (LTRs). Retrotector (Sperber et al., 2007) uses less stringent criteria, therefore false positives must be taken into account. A combination of all three methods may be the best choice to find a balance between sensitivity and specificity. The nonretroviral Penelope family of retrotransposons also contains a distinctive LTR that likely reflects its lack of integrase and dependence on host telomerase.

Retrotransposons without LTRs are called either LINEs or SINEs (Jurka et al., 2007). Long-interspersed nuclear elements (LINEs) in eukaryotes span over 5 kb or greater genomic regions (Gogvadze and Buzdin, 2009), with a RNA polymerase II binding site 5′ of a high-specificity reverse transcriptase gene. Inactivated LINEs are often truncated at the 5′ end. The LINE may also contain an endonuclease with site-specific integration activity. The 3′ end contains a polyadenylation signal (AATAA), usually followed by a G/T-rich region. There are three families of LINEs, based on their origin: in humans, only LINE1 elements are actively transposable, with LINE2 and CR1/LINE3 elements mostly degraded (Gogvadze and Buzdin, 2009).

Short–interspersed nuclear elements (SINEs) are typically up to 500 bases long (Gogvadze and Buzdin, 2009), and are lineage specific. They do not have a protein-coding payload. The most famous of these is the primate-specific *Alu* SINE, which comprises over 10% of the human genome. This SINE is derived from the 7SL RNA, and like most SINEs is a source of recombination events in genome evolution. Because SINE elements are lineage specific, homology search against databases of known SINEs is important. REPBASE (Jurka et al., 2005) is a database with the largest source of known repetitive elements, submitted by researchers and derived from combing the literature. RepeatMasker (www.repeatmasker.org) allows one to use various alignment methods to match against such repeat sequence databases. REPBASE's maintainers also provide a repeat finding system, Censor (Kohany et al., 2006). In either case, sensitivity thresholds play an important role in determining how sensitive repeat masking will be, especially when the organism being masked is distantly related to the well-characterized species from which most REPBASE entries are derived. A database of known repeats is particularly useful for masking sequences during transcript or genome assembly, where there is not enough sequence information available to do *de novo* identification of repeats.

3.5 OTHER REPEATS

De novo methods for repeat identification are also essential, because the databases of known elements are not complete. The task of finding approximate substrings of length n or longer, with less than $x\%$ mismatch is computationally expensive on a genome-wide scale. EulerAlign (Zhang and Waterman, 2005) uses a data structure known as a de Bruijn graph to make the task time approximately linear to the sequence length. REPuter (Kurtz et al., 2001) is able to handle errors or degeneracy in a repeat sequence based on a specialized algorithm, with runtime increasing significantly based on the number of errors to allow. The RepeatModeler software, available from the RepeatMasker Web site (www.repeatmasker.org), uses two different methods of *de novo* prediction and subsequently combines the results. In general, each method has some strengths and weaknesses based on the most important factor to the software author (e.g., speed, sensitivity, accuracy). It is preferable therefore to annotate repetitive elements in a genome based on the concurrence of two or more methods.

Another popular tool for *de novo* repeat detection is PILER (Edgar and Myers, 2005), which uses a unique filtering method for long local alignments. PILER can be used to annotate four types of repeats: dispersed family, tandem array, pseudosatellite, and LTRs. Dispersed families have three or more copies throughout the genome, flanked by unique sequences in each case. Although these are typically genetically mobile elements, they may also be exons from paralogous genes. A tandem array is a contiguous series of three or more copies of a motif within a chosen distance threshold. A pseudo-satellite has clustered copies, but the copies do not meet the distance threshold set for tandem arrays. Short tandem repeats of 1 to 6 base motifs (also known as microsatellites) are most sensitively detected by dedicated software such as SciRoKo (Kofler et al., 2007).

The overall diversity of repetitive elements is reflected in the fact that as of 2011, there are 75 "repeat" related terms in the Sequence Ontology (Eilbeck et al., 2005). The TEclass (Abrusán et al., 2009) software can help annotate the results of *de novo* repeat analyses at the top level of DNA transposon/SINE/LINE/LTR. REPBASE provides a rough classification system for known repeats as well. Beyond these high-level annotations, more refined classification of repetitive elements is not well automated to date.

REFERENCES

Abrusán, G., Grundmann, N., DeMester, L., Makalowski, W. 2009. TEclass—A tool for automated classification of unknown eukaryotic transposable elements. *Bioinformatics* 25:1329–1330.

Aerts, S., Van Loo, P., Thijs, G., et al. 2005. TOUCAN 2: The all-inclusive open source workbench for regulatory sequence analysis. *Nucleic Acids Res.* 33:W393–W396.

Amaral, P.P., Clark, M.B., Gascoigne, D.K. 2011. lncRNAdb: A reference database for long noncoding RNAs. *Nucleic Acids Res.* 39:D146–D151.

Bailey, T.L., Boden, M., Buske, F.A., et al. 2009. MEME SUITE: Tools for motif discovery and searching. *Nucleic Acids Res.* 37:W202–W208.

Balakirev, E.S., Ayala, F.J. 2003. Pseudogenes: Are they "junk" or functional DNA? *Annu. Rev. Genet.* 37:123–151.

Bartsch, D., Casadio, A., Karl, K.A., Serodio, P., Kandel, E.R. 1998. CREB1 encodes a nuclear activator, a repressor, and a cytoplasmic modulator that form a regulatory unit critical for long-term facilitation. *Cell* 95:211–223.

Bengert, P., Dandekar, T. 2004. Riboswitch finder—A tool for identification of riboswitch RNAs. *Nucleic Acids Res.* 32:W154–W159.

Blot, M. 1994. Transposable elements and adaptation of host bacteria. *Genetica* 93:5–12.

Cartharius, K., Frech, K., Grote, K., et al. 2005. MatInspector and beyond: Promoter analysis based on transcription factor binding sites. *Bioinformatics* 21:2933–2942.

Chan, P.P., Cozen, A.E., Lowe, T.M. 2011. Discovery of permuted and recently split transfer RNAs in Archaea. *Genome Biol.* 12:R38.

Chaudhuri, B.N., Yeates, T.O. 2005. A computational method to predict genetically encoded rare amino acids in proteins. *Genome Biol.* 6:R79.

Conlon, E.M., Liu, X.S., Lieb, J.D., Liu, J.S. 2003. Integrating regulatory motif discovery and genome-wide expression analysis. *Proc. Natl. Acad. Sci. USA* 100:3339–3344.

Crooks, G.E., Hon, G., Chandonia, J.M., Brenner, S.E. 2004. WebLogo: A sequence logo generator. *Genome Res.* 14:1188–1190.

Edgar, R.C. 2007. PILER-CR: Fast and accurate identification of CRISPR repeats. *BMC Bioinformatics* 8:18.

Edgar, R.C., Myers, E.W. 2005. PILER: Identification and classification of genomic repeats. *Bioinformatics* 21 (Suppl. 1):i152–i158.

Edlefsen, P.T., Liu, J.S. 2010. Transposon identification using profile HMMs. *BMC Genomics* 11(Suppl. 1):S10.

Eilbeck, K., Lewis, S., Mungall, C.J., et al. 2005. The Sequence Ontology: A tool for the unification of genome annotations. *Genome Biol.* 6:R44.

Feschotte, C., Pritham, E.J. 2007. DNA transposons and the evolution of eukaryotic genomes. *Ann. Rev. Genet.* 41:331–368.

Feschotte, C., Pritham, E. 2009. A cornucopia of Helitrons shapes the maize genome. *Proc. Natl. Acad. Sci. USA* 106 (47):19747–19748.

Freyhult, E., Edvardsson, S., Tamas, I., Moulton, V., Poole, A.M. 2008. Fisher: A program for the detection of H/ACA snoRNAs using MFE secondary structure prediction and comparative genomics—Assessment and update. *BMC Res. Notes* 1:49.

Frith, M.C., Li, M.C., Weng, Z. 2003. Cluster-Buster: Finding dense clusters of motifs in DNA sequences. *Nucleic Acids Res.* 31:3666–3668.

Frith, M.C., Spouge, J.L., Hansen, U., Weng, Z. 2002. Statistical significance of clusters of motifs represented by position specific scoring matrices in nucleotide sequences. *Nucleic Acids Res.* 30(14):3214–3224.

Gardner, P.P., Daub, J., Tate, J., et al. 2011. Rfam: Wikipedia, clans and the "decimal" release. *Nucleic Acids Res.* 39(Database issue):D141–D145.

Gifford, R., Tristem, M. 2003. The evolution, distribution and diversity of endogenous retroviruses. *Virus Genes* 26(3):291–315.

Gogvadze, E., Buzdin, A. 2009. Retroelements and their impact on genome evolution and functioning. *Cell. Mol. Life Sci.* 66:3727–3742.

Grissa, I., Vergnaud, G., Pourcel, C. 2007. The CRISPRdb database and tools to display CRISPRs and to generate dictionaries of spacers and repeats. *BMC Bioinformatics* 8:172.

Grün, D., Wang, Y.L., Langenberger, D., Gunsalus, K.C., Rajewsky, N. 2005. microRNA target predictions across seven *Drosophila* species and comparison to mammalian targets. *PLoS Comput. Biol.* 1:e13.

Hallikas, O., Palin, K., Sinjushina, N., et al. 2006. Genome-wide prediction of mammalian enhancers based on analysis of transcription-factor binding affinity. *Cell* 124:47–59.

Hertel, J., Hofacker, I.L., Stadler, P.F. 2008. SnoReport: Computational identification of snoRNAs with unknown targets. *Bioinformatics* 24:158–164.

Ho Sui, S.J., Fulton, D.L., Arenillas, D.J., Kwon, A.T., Wasserman, W.W. 2007. oPOSSUM: Integrated tools for analysis of regulatory motif over-representation. *Nucleic Acids Res.* 35:W245–W252.

Hu, M., Yu, J., Taylor, J.M., Chinnaiyan, A.M., Qin, Z.S. 2010. On the detection and refinement of transcription factor binding sites using ChIP-Seq data. *Nucleic Acids Res.* 38(7):2154–2167.

Jurka, J., Kapitonov, V.V., Kohany, O., Jurka, M.V. 2007. Repetitive sequences in complex genomes: Structure and evolution. *Annu. Rev. Genomics Hum. Genet.* 8:241–259.

Jurka, J., Kapitonov, V.V., Pavlicek, A., Klonowski, P., Kohany, O., Walichiewicz, J. 2005. Repbase Update, a database of eukaryotic repetitive elements. *Cytogenet. Genome Res.* 110:462–467.

Kapitonov, V.V., Jurka, J. 2006. Self-synthesizing DNA transposons in eukaryotes. *Proc. Natl. Acad. Sci. USA* 103:4540–4545.

Kazazian, H.H. Jr. 1998. Mobile elements and disease. *Curr. Opin. Genet. Dev.* 8:343–350.

Kehr, S., Bartschat, S., Stadler, P.F., Tafer, H. 2011. PLEXY: Efficient target prediction for box C/D snoRNAs. *Bioinformatics* 27:279–280.

Kertesz, M., Iovino, N., Unnerstall, U., Gaul, U., Segal, E. 2007. The role of site accessibility in microRNA target recognition. *Nat. Genet.* 39:1278–1284.

Kiss, T. 2002. Small nucleolar RNAs: An abundant group of noncoding RNAs with diverse cellular functions. *Cell* 109:145–148.

Kofler, R., Schlotterer, C., Lelley, T. 2007. SciRoKo: A new tool for whole genome microsatellite search and investigation. *Bioinformatics* 23:1683–1685.

Kohany, O., Gentles, A.J., Hankus, L. 2006. Annotation, submission and screening of repetitive elements in Repbase: RepbaseSubmitter and Censor. *BMC Bioinformatics* 7:474.

Kulakovskiy, I.V., Boeva, V.A., Favorov, A.V., Makeev, V.J. 2010. Deep and wide digging for binding motifs in ChIP-Seq data. *Bioinformatics* 26(20):2622–2623.

Kurtz, S., Choudhuri, J.V., Ohlebusch, E., Schleiermacher, C., Stoye, J., Giegerich, R. 2001. REPuter: The manifold applications of repeat analysis on a genomic scale. *Nucleic Acids Res.* 29:4633–4642.

Lagesen, K., Hallin, P., Rødland, E.A., Staerfeldt, H.H., Rognes, T., Ussery, D.W. 2007. RNAmmer: Consistent and rapid annotation of ribosomal RNA genes. *Nucleic Acids Res.* 35: 3100–3108.

Laslett, D., Canbäck, B. 2004. ARAGORN, a program to detect tRNA genes and tmRNA genes in nucleotide sequences. *Nucleic Acids Res.* 32:11–16.

Lerat, E., Ochman, H. 2005. Recognizing the pseudogenes in bacterial genomes. *Nucleic Acids Res.* 33(10):3125–3132.

Lewis, B.P., Burge, C.B., Bartel, D.P. 2005. Conserved seed pairing, often flanked by adenosines, indicates that thousands of human genes are microRNA targets. *Cell* 120:15–20.

Liao, P.Y., Choi, Y.S., Lee, K.H. 2009. FSscan: A mechanism-based program to identify +1 ribosomal frameshift hotspots. *Nucleic Acids Res.* 37:7302–7311.

Lowe, T.M., Eddy, S.R. 1997. tRNAscan-SE: A program for improved detection of transfer RNA genes in genomic sequence. *Nucleic Acids Res.* 25:955–964.

Lowe, T.M., Eddy, S.R. 1999. A computational screen for methylation guide snoR-NAs in yeast. *Science* 283:1168–1171.

Matys, V., Kel-Margoulis, O.V., Fricke, E., et al. 2006. TRANSFAC and its module TRANSCompel: Transcriptional gene regulation in eukaryotes. *Nucleic Acids Res.* 34:D108–D110.

McCarthy, E., McDonald, J. 2003. LTR_STRUC: A novel search and identification program for LTR retrotransposons. *Bioinformatics* 19:362–367.

Pavesi, G., Mereghetti, P., Mauri, G., Pesole, G. 2004. Weeder Web: Discovery of transcription factor binding sites in a set of sequences from co-regulated genes. *Nucleic Acids Res.* 32:W199–W203.

Portales-Casamar, E., Kirov, S., Lim, J. et al. 2007. PAZAR: A framework for collection and dissemination of cis-regulatory sequence annotation. *Genome Biol.* 8:R207.

Reifur, L., Yu, L.E., Cruz-Reyes, J., Vanhartesvelt, M., Koslowsky, D.J. 2010. The impact of mRNA structure on guide RNA targeting in kinetoplastid RNA editing. *PLoS ONE* 5(8):e12235.

Ro, S., Park, C., Song, R., et al. 2007. Cloning and expression profiling of testis-expressed piRNA-like RNAs. *RNA* 13:1693–1702.

Sai Lakshmi, S., Agrawal, S. 2008. piRNABank: A Web resource on classified and clustered Piwi-interacting RNAs. *Nucleic Acids Res.* 36:D173–D177.

Sandelin, A., Wasserman, W.W. 2004. Constrained binding site diversity within families of transcription factors enhances pattern discovery bioinformatics. *J. Mol. Biol.* 338:207–215.

Sandelin, A., Wasserman, W.W., Lenhard, B. 2004. ConSite: Web-based prediction of regulatory elements using cross-species comparison. *Nucleic Acids Res.* 32:W249–W252.

Schattner, P., Decatur, W.A., Davis, C., Ares, M. Jr., Fournier, M.J., Lowe, T.M. 2004. Genome-wide searching for pseudouridylation guide snoRNAs: Analysis of the *Saccharomyces cerevisiae* genome. *Nucleic Acids Res.* 32:4281–4296.

Schmid, C.D., Perier, R., Praz, V., Bucher, P. 2006. EPD in its twentieth year: Towards complete promoter coverage of selected model organisms. *Nucleic Acids Res.* 34:D82–D85.

Shimizu, T.S., Takahashi, K., Tomita, M. 1997. CpG distribution patterns in methylated and non-methylated species. *Gene* 205:103–107.

Silva, J.M., Sachidanandam, R., Hannon, G.J. 2003. Free energy lights the path toward more effective RNAi. *Nat. Genet.* 35:303–305.

Sperber, G.O., Airola, T., Jern, P., Blomberg, J. 2007. Automated recognition of retroviral sequences in genomic data—RetroTector. *Nucleic Acids Res.* 35:4964–4976.

Sugahara, J., Yachie, N., Sekine, Y., et al. 2006. SPLITS: A new program for predicting split and intron-containing tRNA genes at the genome level. *In Silico Biol.* 6:411–418.

Tafer, H., Kehr, S., Hertel, J. 2010. RNAsnoop: Efficient target prediction for H/ACA snoRNAs. *Bioinformatics* 26:610–616.

Theis, C., Reeder, J., Giegerich, R. 2008. KnotInFrame: Prediction of –1 ribosomal frameshift events. *Nucleic Acids Res.* 36:6013–6020.

Thomas-Chollier, M., Sand, O., Turatsinze, J.V., et al. 2008. RSAT: Regulatory sequence analysis tools. *Nucleic Acids Res.* 36:W119–W127.

Volff, J.N. 2006. Turning junk into gold: Domestication of transposable elements and the creation of new genes in eukaryotes. *Bioessays* 28:913–922.

Xie, X., Lu, J., Kulbokas, E.J., et al. 2005. Systematic discovery of regulatory motifs in human promoters and 3′ UTRs by comparison of several mammals. *Nature* 434:338–345.

Yan, Z., Hu, H.Y., Jiang, X. et al. 2011. Widespread expression of piRNA-like molecules in somatic tissues. *Nucleic Acids Res.* 39:6596–6607.

Yang, J.H., Zhang, X.C., Huang, Z.P., et al. 2006. snoSeeker: An advanced computational package for screening of guide and orphan snoRNA genes in the human genome. *Nucleic Acids Res.* 34:5112–5123.

Yano, Y., Saito, R., Yoshida, N., et al. 2004. A new role for expressed pseudogenes as ncRNA: Regulation of mRNA stability of its homologous coding gene. *J. Mol. Med.* 82:414–422.

Zhang, Y., Waterman, M.S. 2005. An Eulerian path approach to local multiple alignment for DNA sequences. *Proc. Natl. Acad. Sci. USA* 102:1285–1290.

Zhang, Z., Harrison, P., Gerstein, M. 2002. Identification and analysis of over 2000 ribosomal protein pseudogenes in the human genome. *Genome Res.* 12(10):1466–1482.

Zhou, Q., Wong, W.H. 2004. CisModule: *De novo* discovery of cis-regulatory modules by hierarchical mixture modeling. *Proc. Natl. Acad. Sci. USA* 101(33):12114–12119.

CHAPTER 4

Genome-Associated Data

4.1 INTRODUCTION

Determining the location of genes alone is of course not sufficient to get a comprehensive picture of the dynamic process called life. If genes are the program for an organism's life, then gene expression is the equivalent of the running software process. Transcription factor analysis (see Chapter 3) is just the beginning of the process used to determine how genes work together and how a functioning organism works overall. In prokaryotes, expression in neighboring genes in the genome can be coordinated (through cotranscription in operons, for instance). In higher eukaryotes, the burgeoning field of epigenetics attests to the additional levels of gene expression control implemented in these organisms by its name: *epi-* means "outside" or "above" in the Greek language. Adding another level of complexity, the concept of "the" genome for a "species" is becoming more or less obsolete. High-throughput sequencing has facilitated the elucidation of many *individual* human genomes. At the same time, sequencing uncultured prokaryotes has revealed that delineating species by sequence data is nontrivial, due to such factors as widespread lateral gene transfer (Boucher et al., 2003). While the tools for analyzing genomic data are quickly evolving, this chapter provides a brief introduction to these new data annotation concepts and extant tools.

4.2 OPERONS

In prokaryotes, multiple neighboring genes on the same strand can be cotranscribed; such a genomic region is called an operon. The genes in an operon are normally involved in the same biological process, such as a metabolic pathway. If all but one of the genes coding for a reaction cascade are in an operon and a gene with unknown function is also found in the same operon, this location information can be used as circumstantial evidence that the gene with unknown function may perform the missing reaction. This can be a lead for follow-up biochemical validation. This form of functional annotation is independent of the other annotation techniques described in Chapter 7.

Prediction of operon structure has traditionally been based on observing smaller-than-normal intergene gaps. Cotranscription also implies only one ribosomal binding site (Shine–Dalgarno pattern from Chapter 2) associated with the most 5′ gene, and often a stem–loop structure to terminate transcription at the 3′ end. Relatively high false-positive and false-negative rates for these signals (which are very short patterns) mean that other techniques need to be used to improve the accuracy of operon prediction. Predictions can be measured against data such as coexpression analysis (e.g., microarrays) or RNA evidence. The most accurate extant operon prediction software (~94%), called STRING (Taboada et al., 2010), uses intergenic distance, plus functional clustering of genes based on protein interaction data.

4.3 METAGENOMICS

Metagenomics is the study of genomic DNA extracted from a mixed population of microbial species, typically directly from the environment rather than cultured in media in a lab. The principal purpose of such a data collection is to gain an understanding of the dynamics of microbial ecosystems in nature. For this reason, this technique is also known as ecogenomics (Ouborg and Vriezen, 2007). The analysis and annotation of this mixed data requires several additional considerations: population statistics, data size, phylogenetic sorting, and quality of the assembly.

4.3.1 Population Statistics

Many of the sample mixture models used in metagenomics data analysis are naturally borrowed from and are reshaping ecology, which has an established tradition of population statistics (Waples and Gaggiotti, 2006). Building on multiple existing programs, one of the most comprehensive tools for analyzing genetic data in ecology is mothur (Schloss et al., 2009). A fundamental concept for sequence-based analysis of microbial populations is setting aside the strict idea of distinct species, and instead clustering variable (but closely related) genomic data into so-called operational taxonomic units (OTUs). Typically the 16S ribosomal RNA is selected for OTU analysis, since universal primers targeting highly conserved regions of the rRNA can be used to amplify this gene across the entire microbial population in the sample. After sequence generation and clustering, the same OTU is normally assigned to all sequences with either less than 3% or 5% sequence variance. The methods for calculating this variance (i.e., distance) are traditionally based on initially performing a multiple sequence alignment of the whole data set. This ensures that variance measures are consistent across all possible pairs of sequences. This is the method used for instance by mothur. An alternative would be to simply perform pairwise alignments instead of the multiple alignment, but this typically leads to an underestimate of the variance among all of the sequences in an OTU. A compromise is to perform global (i.e., full-length query and target) pairwise alignments. This approach has been reported with favorable results by the authors of the 16S rRNA analysis tool ESPRIT (Sun et al., 2009).

Once a distance measure is established, the clustering of sequences using the calculated distances can be done. From least to most robust, common clustering methods are:

- Single linkage—If sequence *A* is close to *B*, and *B* is close to *C*, then assume *A* and *C* are also close enough to be in the same OTU.

- Centroid linkage—If sequence *T* from an OTU is roughly equidistant from all others in the OTU (i.e., it is the centroid) and sequence *A* is close to *T*, then assume *A* is close enough to all others in the OTU to belong. The "average linkage" method uses a similar process but calculates a statistical average among cluster members instead of using a specific centroid.

- Complete linkage—Sequence *A* is only included in an OTU if it is close to all members already in the OTU.

For an overview of clustering techniques, consult Felsenstein (2004). Once OTUs for a sample are established, many statistical tests can be applied to the OTU frequency data. For example, within a sample, the saturation of the data set can be estimated (i.e., will generating more sequence depth through additional DNA sequencing help to uncover more OTUs?). The diversity of the sample can also be evaluated using classic information theory techniques such as Shannon entropy (disorder) (Lin, 1991).

Intra-sample comparisons of OTU distribution can be made to evaluate the similarity of multiple microbial environments.

4.3.2 Data Size

With the advent of high-throughput metagenomics, the traditional and robust methods for 16S distance and OTU calculation can prove too computationally costly for even large bioinformatics facilities. At the cost of some accuracy, QIIME (Caporaso et al., 2010) can be used to rapidly calculate OTUs based on pairwise alignments and single linkage. Removing exact duplicates and shorter subsequences from the initial data set can simplify the OTU creation task, as long as measures are in place to include these initially removed data in the final OTU frequency counts (through a mapping step, which links these data with the overall set of clusters). Another approach is to divide the initial sequence data set into large superclusters (e.g., all sequences within 30% distance), using a tool originally intended for transcript-sequence clustering such as CD-HIT (Niu et al., 2010). These superclusters can subsequently be processed with robust methods (and in parallel) without the risk of significantly affecting the final OTU assignments. A case study in metagenomics analysis methodology is presented in Chapter 11.

It is worth noting that during the 16S RNA sequencing, stringent quality control (e.g., clipping) should be applied to the incoming data set, because the individual reads cannot be assembled together and need to be analyzed individually. With large data sets in particular, there is also a significant risk that chimeric sequences (i.e., sequences artificially created by coligation of disparate sequence fragments) are likely to inflate the total number of OTUs detected. Chimeric sequence checking algorithms, such as UCHIME (Edgar et al., 2011), can be used to filter these out.

4.3.3 Phylogenetic Sorting

Once OTUs are established and population statistics calculated, insight into the possible ecological roles of the constituent organisms can be roughly inferred by linking the OTUs to the phylogenetic tree. Once the taxonomic distribution is available, one might see, for example, that families of sulfur-reducing microbes are prevalent in a particular sample. Traditionally, a phylogenetic tree would be constructed for related sequences, for example, using distance, parsimony, or maximum likelihood methods in the MEGA software (Tamura et al., 2011). Unfortunately, metagenomic data sets are simply too large to use in regular tree construction methods.

Specific computational methods for assigning OTUs taxonomy labels are varied, with the popular methods listed roughly from fastest to slowest:

- Bayesian classifiers trained on a reference set—The Ribosomal Database Project (RDP) (Wang et al., 2007).

- Alignment to reference 16S multiple sequence alignment—Greengenes (DeSantis et al., 2006) or Silva (Pruesse et al., 2007) reference, using the ARB taxonomy software (Ludwig et al., 2004).

- Parsimonious tree insertion—Also available from ARB (Ludwig et al., 2004).

Reporting of the taxonomy to the user can be useful at different levels, depending on their interest: phylum, class, order, family, or genus. A caveat about such assignments is that different taxonomy databases exist within all of the aforementioned software, and therefore the resulting labels can differ, even when the same sequence data is used. Another caveat is that a growing proportion of the reference 16S RNA sequence data set possesses only loosely assigned labels (e.g., "uncultured Sulfolobales"), as metagenomic data sets are submitted to the public databases with very little annotation.

Although not as accurate, a method applicable to both 16S RNA and whole-genomic metagenomic taxonomy assignments is pairwise alignment (e.g., BLAST) to a reference database. Strong alignments can be analyzed for their taxonomic distribution, and a consensus label can be inferred. This method is used for example by the MEGAN software (Huson et al., 2007).

4.3.4 Assembly Quality

Whole metagenomic samples pose some additional challenges to genome annotation. In many cases, only genomic fragments can be assembled, ranging from single reads to fragments with a size of a few hundred kilobase pairs. Coding sequence prediction should be done using tools that do not require training sets, since the different OTUs in the sample may not have the same coding sequence model; they belong to a large number of species, possibly even belonging to different parts of the phylogenetic tree. Another option is to put the assembled contigs into taxonomic "bins" in a manner akin to OTU assignment. This can be done using BLAST searches or statistical techniques, such as classifiers based on N-mer frequency distributions (McHardy et al., 2007). When 16S data for the same sample is available, concurrence of the genomic bin and the 16S OTU distributions can be used as a form of assembly quality control.

It is also quite likely that a metagenomic assembly will contain at least some chimeric contigs. While phylogenetic analysis of the coding genes may indicate a taxonomic boundary in the contig, for smaller contigs this is difficult to distinguish from an instance of lateral gene transfer. One of the more complete genome annotation pipelines dedicated to metagenomic data is the Joint Genome Institute's IMG/M (Markowitz et al., 2008).

4.4 INDIVIDUAL GENOMES

4.4.1 Epigenetics

4.4.1.1 DNA Methylation

Methylation of cytosine bases in the genome occurs mainly at CpG islands (see Chapter 3). These islands overlap frequently with transcription factor binding sites, and methylation has the net effect of suppressing gene expression. While methylation patterns are inheritable, overall methylation patterns may vary even between genetically identical siblings (Petronis, 2006) and between different cell types (which require different levels of transcriptional activity) in an individual (Illingworth et al., 2008). Sequencing of bisulfite-treated DNA yields dense read coverage near methylation sites (Ku et al., 2011), and various statistics-intensive computational methods can be used to infer the methylation patterns. While there is not yet a standard file representation for DNA methylation data, tools such as the BDCP Web server (Rohde et al., 2008) can be used to generate data files in GFF3 format (www.sequenceontology.org/gff3. shtml). These can be used as custom tracks in the University of California, Santa Cruz (UCSC) genome browser (see Chapter 9) and also many other genome viewing tools. Due to the nature of the methylation phenomenon, public databases for methylation data (e.g., Negre and Grunau, 2006) are broken down by species, tissues, cell lines, and phenotypes.

4.4.1.2 Histone Modification

Various histone proteins heterodimerize in order to form the nucleosome (Park and Luger, 2006). Approximately 142 to 152 bases of genomic DNA wrap around the nucleosome, and nucleosomes are separated by about 50 bases of free DNA bound only to histone H1. Histone proteins are

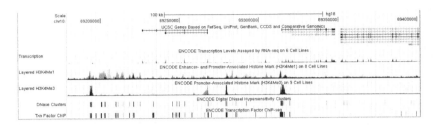

FIGURE 4.1 Visualization of quantitative epigenetic evidence in the UCSC Genome Browser. Peaks represent areas that are likely to be regulated via histone modification.

subject to many chemical modifications, which either promote or suppress transcription. The genome portion, where the histone modifications make it "open" and transcriptionally active, is called euchromatin (Babu and Verma, 1987). Suppressing modifications cause the formation of tightly packed histone arrays, which are called heterochromatin (Babu and Verma, 1987). Euchromatic profiling for whole genomes is a relatively new field and poses many computational challenges. As these features are fairly large and discrete (on/off), the widely supported BED and BigBED file formats (Kent et al., 2010) are common representations for this type of data. These data are also available from public databases such as NCBI GEO (Barrett et al., 2011). The ENCODE Project Consortium (2011) has made available large amounts of quantitative epigenetic data, as well as postanalysis to call discrete peak regions. These data include mapped sequencing depth of chromatin immunoprecipitated sequence (ChIP-Seq), as in Figure 4.1.

4.4.1.3 Nucleosome Positioning

The position of nucleosomes in a coding region can affect transcript levels via interference with transcription initiators (positions near the start site) and transcription elongation factors (positions downstream) (Pokholok et al., 2005). Large-scale physical elucidation of nucleosome positioning and modification in smaller eukaryotic genomes has been published (Pokholok et al., 2005; Westenberger et al., 2009), including changes during organism life cycle. Bioinformatics approaches to nucleosome position prediction (e.g., Segal et al., 2006) generally revolve around the ease with which a given DNA segment bends, since this correlates with propensity to wrap around a nucleosome. Results are typically stored as GFF3 files.

4.4.2 Single Nucleotide Polymorphisms

4.4.2.1 Nomenclature

Single nucleotide polymorphisms (SNPs, pronounced "snips") are single bases that differ in an individual relative to a reference genome. For any individual human, 0.1% of the genome constitutes SNPs (Pushkarev et al., 2009), a relatively low number compared to most species. Within the coding sequence regions, this drops to 0.048%, with roughly 37% of those SNPs being homozygous (Pelak et al., 2010). Reporting an entire individual sequence is highly redundant, because of the relatively small number of differences from the known reference. It is therefore much more relevant to annotate only the variants, using a common syntax. Variant calling software tools typically report in the Variant Call Format (VCF) devised by the 1000 Genomes Project (Danecek et al., 2001). These files are succinct but fairly impenetrable for human consumption.

For legible reporting, the Human Genetic Variation Society has devised a standardized annotation shorthand (HGVS syntax) for describing SNPs and other sequence variants (den Dunnen and Antonarakis, 2000). In a coding region, a SNP can be described relative to the start of the coding sequence. For example, BRCA1:c.100A>G indicates that base 100 of the coding sequence for the BRCA1 gene has A as the reference base, and the called variant in an individual genome is G. When the gene context is clear, the gene symbol from the HUGO Gene Nomenclature Committee (Seal et al., 2011) can be omitted. Variants in the 5′ untranslated region (UTR) are described using negative positions; c.−15T>G indicates a SNP 15 bases upstream of the start codon (there is no position 0). Variants after the stop codon are indicated with an asterisk; c.*22T>C indicates a SNP 22 bases downstream of the third base in the stop codon. Variants in introns are indicated as either downstream of the donor site position (e.g., c.256+2T>C is a variant in the second base of the donor site) or upstream of the acceptor site (e.g., c.257−20G>T is a variant 20 bases upstream of the acceptor site, or 18 bases upstream of the "AG" acceptor sequence).

4.4.2.2 Effects

The coding sequence effect of a SNP can generally be classified as:

- Silent/synonymous—In a coding region, the base change does not affect the amino acid translation of the gene. While these changes are generally ignored, it has been noted in prokaryotes that either translation efficiency can be affected by codon usage (Plotkin and Kudla, 2011).

- Missense/nonsynonymous—The amino acid translation will be different from that in the reference genome.

- Nonsense—The SNP introduces an in-frame stop codon.

- Read through—The SNP removes the reference stop codon.

The HGVS syntax allows for the description of the variant at the protein level as well. A change of glutamine to arginine at residue position 488 would be "p.Gln488Arg". Single amino acid IUPAC codes (Liébecq, 1992) can also be used. SNPs can affect the protein translation in other ways as well. A SNP introducing a premature stop codon would be noted as "p.Gln488Ter". Read through has the syntax "p.Ter797Cys*25" if the change creates a cysteine and the next in-frame stop is 25 residues (i.e., 75 bases) downstream.

Some SNP variants located in exons may create new ("cryptic") acceptor or donor splice sites (e.g., Dunn et al., 2002). SNP variants in introns may also disrupt the donor or acceptor site. They may also, in rare instances, create a small novel exon inside of what would normally be an intron (also known as a cryptic exon). In the 5′ UTR, a SNP may disrupt the binding of a transcription factor or create a new preferred start codon. In the 3′ UTR a SNP may disrupt the polyadenylation signal for a gene, shortening its half-life (Kwan et al., 2008). A 3′ UTR SNP may also disrupt a microRNA suppression mechanism, and these are starting to be catalogued (Bruno et al., 2012). All of these possibilities point to the importance of correct annotation of genes and associated control elements.

VCF files from the 1000 Genomes Project are publically available. SNP variants and some small insertions or deletions (indels) reported in the literature are available from the NCBI's dbSNP database (Day, 2010), with links to other resources such as information about related diseases.

4.4.3 Insertions and Deletions

4.4.3.1 Nomenclature

Insertions are extra bases (as opposed to base substitutions) that exist in the individual but not in the reference genome. These are also reportable in VCF files. The HGVS syntax is "c.22_23insTTG" for the insertion of TTG between bases 22 and 23 of the coding sequence. Deletions are the removal of bases in the individual, relative to the reference genome. The HGVS syntax is "c.22_26del" for the deletion of five (end–start+1) bases, or in the case of a single-base deletion the syntax "c.22del" is allowed. The sequence, which is deleted in comparison to the reference sequence, can optionally be appended to this syntax (e.g., c.11delT).

Complex variants composed of simultaneous insertion and deletion events are often referred to as indels. The HGVS syntax is "c.345_347delinsTTTG" for the replacement of three bases in the reference sequence by the four bases TTTG. A special case of an indel is an inversion, where the insertion sequence is of the same length as the deletion, and the two sequences are reverse complementary. The HGVS syntax for this case is "c.665_670inv" for a six-base inversion.

Because they span multiple reference base positions, insertions and deletions can be quite complex in their nomenclature. For example, "c.344_347+2del" is a six-base deletion, removing four coding bases and the (presumable) intron donor GT. The syntax "c.*89-7_*89-3delinsAA-GAA" is a five-base indel, just before the acceptor site of a 3′ UTR exon. A special case of insertions is duplication of short stretches of sequence from the reference genome, with the syntax "c.345_348dup".

4.4.3.2 Effects

In addition to causing the same effects as SNPs, insertions and deletions may case the insertion or deletion of amino acids in the respective protein. If the variant lengths are not multiples of three, translation frameshifts will also occur. As an example, the HGVS syntax is "p.Cys254Trpfs*39" if the frameshift causes an amino acid change from cysteine to tryptophan, and the new in-frame stop is 39 residues away. Listing the new stop location is optional, making "p.Cys254Trpfs" also valid.

4.4.4 Copy Number Variation

In polyploid genomes, variants in the number of copies of a chromosomal segment occur in individuals (Freeman et al., 2006). These copy number variants (CNVs) may affect the organism via gene dosage effects and can lead to disease (Almal and Padh, 2012). Methods for high-throughput sequencing of CNV calls are nascent, with most CNV data being generated using comparative genomic hybridization (CGH) microarrays (Alkan et al., 2011). These methods can be used to generate a relative copy number estimate (e.g., –1 or 1) over a region defined by the density of probes on the microarray.

Where high-resolution sequence data is available, the Wiggle file format (Kent et al., 2010) is suitable for storing and subsequently plotting the data in the UCSC genome browser and others (see Chapter 9). Where CNV data is more coarse-grained, the BedGraph file format (Kent et al., 2010) can succinctly represent the information and is accepted by many of the same browsers.

Public repositories for CNVs, large insertions and deletions, and translocations (i.e., structural variants), called dbVAR and DGVa, are now being coordinated by an international consortium (Church et al., 2010). Cancer is generally the product of *de novo* mutations in particular genes, or the gain or loss of larger chromosomal segments. An excellent public resource for CGH information focused on cancer-related structural changes is Progenetix (Baudis, 2006).

REFERENCES

Alkan, C., Coe, B.P., Eichler, E.E. 2011. Genome structural variation discovery and genotyping. *Nat. Rev. Genet.* 12(5):363–376.

Almal, S.H., Padh, H. 2012. Implications of gene copy-number variation in health and diseases. *J. Hum. Genet.* 57(1):6–13.

Babu, A., Verma, R.S. 1987. Chromosome structure: Euchromatin and heterochromatin. *Int. Rev. Cytol.* 108:1–60.

Barrett, T., Troup, D.B., Wilhite, S.E., et al. 2011. NCBI GEO: Archive for functional genomics data sets—10 years on. *Nucleic Acids Res.* 39:D1005–D1010.

Baudis, M. 2006. Online database and bioinformatics toolbox to support data mining in cancer cytogenetics. *Biotechniques* 40:296–272.

Boucher, Y., Douady, C.J., Papke, R.T., et al. 2003. Lateral gene transfer and the origins of prokaryotic groups. *Annu. Rev. Genet.* 37:283–328.

Bruno, A.E., Li, L., Kalabus, J.L., Yu, A., Hu, Z. 2012. miRdSNP: A database of disease-associated SNPs and microRNA target sites on 3′ UTRs of human genes. *BMC Genomics* 13(1):44.

Caporaso, J.G., Kuczynski, J., Stombaugh, J., et al. 2010. QIIME allows analysis of high-throughput community sequencing data. *Nat. Methods* 7(5):335–336.

Church, D.M., Lappalainen, I., Sneddon, T.P. 2010. Public data archives for genomic structural variation. *Nat. Genet.* 42(10):813–814.

Danecek, P., Auton, A., Abecasis, G., et al. 2001. The variant call format and VCFtools. *Bioinformatics* 27(15):2156–2158.

Day, I.N. 2010. dbSNP in the detail and copy number complexities. *Hum. Mutat.* 31(1):2–4.

den Dunnen, J.T., Antonarakis, S.E. 2000. Mutation nomenclature extensions and suggestions to describe complex mutations: A discussion. *Hum. Mutat.* 15(1):7–12.

DeSantis, T.Z., Hugenholtz, P., Larsen, N., et al. 2006. Greengenes, a chimera-checked 16S rRNA gene database and workbench compatible with ARB. *Appl. Environ. Microbiol.* 72(7):5069–5072.

Dunn, D.M., Ishigami, T., Pankow, J. et al. 2002. Common variant of human NEDD4L activates a cryptic splice site to form a frameshifted transcript. *J. Hum. Genet.* 47(12):665–676.

Edgar, R.C., Haas, B.J., Clemente, J.C., Quince, C., Knight, R. 2011. UCHIME improves sensitivity and speed of chimera detection. *Bioinformatics* 27(16):2194–2200.

ENCODE Project Consortium, Myers, R.M., Stamatoyannopoulos, J., et al. 2011. A user's guide to the encyclopedia of DNA elements (ENCODE). *PLoS Biol.* 9(4):e1001046.

Felsenstein, J. 2004. *Inferring Phylogenies*. Sunderland, MA: Sinauer Associates, Inc.

Freeman, J.L., Perry, G.H., Feuk, L., et al. 2006. Copy number variation: New insights in genome diversity. *Genome Res.* 16(8):949–961.

Huson, D.H., Auch, A.F., Qi, J., Schuster, S.C. 2007. MEGAN analysis of metagenomic data. *Genome Res.* 17(3):377–386.

Illingworth, R., Kerr, A., Desousa, D., et al. 2008. A novel CpG island set identifies tissue-specific methylation at developmental gene loci. *PLoS Biol.* 6(1):e22.

Kent, W.J., Zweig, A.S., Barber, G., Hinrichs, A.S., Karolchik, D. 2010. BigWig and BigBed: Enabling browsing of large distributed datasets. *Bioinformatics* 26(17):2204–2207.

Ku, C.S., Naidoo, N., Wu, M., Soong, R. 2011. Studying the epigenome using next generation sequencing. *J. Med. Genet.* 48(11):721–730.

Kwan, T., Benovoy, D., Dias, C., et al. 2008. Genome-wide analysis of transcript isoform variation in humans. *Nat. Genet.* 40(2):225–231.

Liébecq, C. 1992. *Biochemical Nomenclature and Related Documents*, 2nd edition. London: Portland Press, pp. 122–126.

Lin, J. 1991. Divergence measures based on the Shannon entropy. *IEEE T. Inform. Theory* 37(1):145–151.

Ludwig, W., Strunk, O., Westram, R., et al. 2004. ARB: A software environment for sequence data. *Nucleic Acids Res.* 32(4):1363–1371.

Markowitz, V.M., Ivanova, N., Szeto, E., et al. 2008. IMG/M: A data management and analysis system for metagenomes. *Nucleic Acids Res.* 36: D534–D538.

McHardy, A.C., Martín, H.G., Tsirigos, A., Hugenholtz, P., Rigoutsos, I. 2007. Accurate phylogenetic classification of variable-length DNA fragments. *Nat. Methods* 4(1):63–72.

Negre, V., Grunau, C. 2006. The MethDB DAS server: Adding an epigenetic information layer to the human genome. *Epigenetics* 1(2):101–105.

Niu, B., Fu, L., Sun, S., Li, W. 2010. Artificial and natural duplicates in pyrosequencing reads of metagenomic data. *BMC Bioinformatics* 11:187.

Ouborg, N.J., Vriezen, W.H. 2007. An ecologist's guide to ecogenomics. *J. Ecol.* 95:8–16.

Park, Y.J., Luger, K. 2006. Structure and function of nucleosome assembly proteins. *Biochem. Cell Biol.* 84(4):549–558.

Pelak, K., Shianna, K.V., Ge, D., et al. 2010. The characterization of twenty sequenced human genomes. *PLoS Genet.* 6(9):e1001111.

Petronis, A. 2006. Epigenetics and twins: Three variations on the theme. *Trends Genet.* 22(7):347–350.

Plotkin, J.B., Kudla, G. 2011. Synonymous but not the same: The causes and consequences of codon bias. *Nat. Rev. Genet.* 12(1):32–42.

Pokholok, D.K., Harbison, C.T., Levine, S., et al. 2005. Genome-wide map of nucleosome acetylation and methylation in yeast. *Cell* 122(4):517–527.

Pruesse, E., Quast, C., Knittel, K., et al. 2007. SILVA: A comprehensive online resource for quality checked and aligned ribosomal RNA sequence data compatible with ARB. *Nucleic Acids Res.* 35(21):7188–7196.

Pushkarev, D., Neff, N.F., Quake, S.R. 2009. Single-molecule sequencing of an individual human genome. *Nat. Biotechnol.* 27(9):847–850.

Rohde, C., Zhang, Y., Jurkowski, T.P., Stamerjohanns, H., Reinhardt, R., Jeltsch, A. 2008. Bisulfite sequencing Data Presentation and Compilation (BDPC) Web server—A useful tool for DNA methylation analysis. *Nucleic Acids Res.* 36(5):e34.

Schloss, P.D., Westcott, S.L., Ryabin, T., et al. 2009. Introducing mothur: Open-source, platform-independent, community-supported software for describing and comparing microbial communities. *Appl. Environ. Microbiol.* 75(23):7537–7541.

Seal, R.L., Gordon, S.M., Lush, M.J., Wright, M.W., Bruford, E.A. 2011. Genenames. org: The HGNC resources in 2011. *Nucleic Acids Res.* 39(Database issue): D514–D519.

Segal, E., Fondufe-Mittendorf, Y., Chen, L., et al. 2006. A genomic code for nucleosome positioning. *Nature* 442(7104):772–778.

Sun, Y., Cai, Y., Liu, L., et al. 2009. ESPRIT: Estimating species richness using large collections of 16S rRNA pyrosequences. *Nucleic Acids Res.* 37(10):e76.

Taboada, B., Verde, C., Merino, E. 2010. High accuracy operon prediction method based on STRING database scores. *Nucleic Acids Res.* 38(12):e130.

Tamura, K., Peterson, D., Peterson, N., Stecher, G., Nei, M., Kumar, S. 2011. MEGA5: Molecular evolutionary genetics analysis using maximum likelihood, evolutionary distance, and maximum parsimony methods. *Mol. Biol. Evol.* 28(10):2731–2739.

Wang, Q., Garrity, G.M., Tiedje, J.M., Cole, J.R. 2007. Naive Bayesian classifier for rapid assignment of rRNA sequences into the new bacterial taxonomy. *Appl. Environ. Microbiol.* 73(16):5261–5267.

Waples, R.S., Gaggiotti, O. 2006. What is a population? An empirical evaluation of some genetic methods for identifying the number of gene pools and their degree of connectivity. *Mol. Ecol.* 15:1419–1439.

Westenberger, S.J., Cui, L., Dharia, N., Winzeler, E., Cui, L. 2009. Genome-wide nucleosome mapping of *Plasmodium falciparum* reveals histone-rich coding and histone-poor intergenic regions and chromatin remodeling of core and subtelomeric genes. *BMC Genomics* 10:610.

CHAPTER 5

Characterization of Gene Function through Bioinformatics

The Early Days

5.1 OVERVIEW

The characterization of gene sequences began with the invention of protein sequencing techniques (Edman, 1949) and intensified after DNA sequencing became a reality (Gilbert and Maxam, 1973; Sanger et al., 1973). In the early days of sequence production, computation was in its infancy, with punch cards and primitive mainframe computers as the only means to perform bioinformatics tasks. Despite these rudimentary environments, researchers began to develop tools that allowed the comparison of sequences (Doolittle, 1981). The next phase of computer development introduced the command line terminals and large-scale data storage on disk drives. Still, the Web did not exist and computing was only accessible to a select minority. With the creation of early personal computers, especially the IBM PC, the Apple IIe, and the Atari 520, this changed dramatically, as all of a sudden access to computers became a possibility for most scientists. At the same time, the programming tools evolved rapidly, so that it was possible for many to create scripts and programs to assist in the analysis of sequence information. To this day, some of the bioinformatics tools developed early on are still in use and students at the university level are still learning how to use a command line interface (now of course on UNIX machines).

It was recognized early on that the newly created protein and DNA sequences needed to be stored in a public repository, similar to books and other written works in libraries; therefore GenBank, the European Molecular Biology Laboratory (EMBL) Data library, and Japan International Protein Information Database (JIPID) were founded (Benson et al., 1993; Kneale and Kennard, 1984). As the Internet really did not yet exist as a commodity in these early days, the data were initially distributed on tapes and floppy disks at a cost. Smith and Waterman created the original exhaustive search algorithm for sequence data in 1984 (Lipman et al., 1984). It turned out that with the computer infrastructure of the time, this algorithm took too long for it to be useful; therefore the first "workable" tools to emerge in bioinformatics were the heuristic database search algorithms FASTA (Pearson and Lipman, 1988) and soon thereafter BLAST (Altschul et al., 1990). These could be used to screen a new sequence against the sequences in the databanks. Specialized search algorithms emerged, which were capable of identifying motifs within protein sequences (such as Prosearch) (Kolakowski et al., 1992) or finding restriction sites within DNA sequences (such as REBASE) (Roberts and Macelis, 1993). Tools for the analysis of biophysical parameters of protein sequences were created. For instance, it became very popular to publish hydrophobicity analyses (Bigelow, 1967) for newly sequenced proteins in the scientific literature. Hydrophobicity plots can also be considered one of the first graphical visualizations used in bioinformatics.

Eventually, the early stand-alone bioinformatics tools were bundled into sequence analysis suites, which were commercially marketed. As in most emerging fields, initially quite a variety of analysis suites existed, from IBI Pustell (Pustell and Kafatos, 1984) over PC/GENE (Moore et al., 1988) to UW GCG (Devereux et al., 1984). While the first two examples mentioned worked on personal computers, the last one (which over time was renamed to "GCG") worked originally on the Digital Equipment (DEC) VMS platform. GCG grew into a fairly large program suite over time and became one of the most used toolkits in bioinformatics, for example, the EMBnet nodes (http://embnet.org), which provided access to bioinformatics resources, standardized on this package worldwide for quite some time.

Initially, the sequences created in the laboratories were only a few amino acids long, and even the early DNA sequences almost never exceeded

1000 bp. One should not forget that it took several months to create even a 1000 bp sequence in the early days of DNA sequencing. As the main memory on personal computers, and even mainframes, was very limited initially (640 kilobytes was considered very large), the bioinformatics tool developers usually restricted the maximum size of a sequence that could be handled at once. For example, sequences in the GCG package were limited to 100,000 amino acids or base pairs. These early restrictions were unfortunately hardcoded, making it very difficult to modify these applications later on to enable the handling of complete genomes, which, even in the case of complete microbial genomes, were of course much larger than this artificial cutoff. When the first complete microbial genomes were first presented in the year 1995, the only tools to handle the complete sequence were word processors, such as Microsoft Word or Wordstar. Once complete genome sequences existed, this limitation was overcome quickly through patches to existing programs and the creation of new software tools.

5.2 STAND-ALONE TOOLS AND TOOLS FOR THE EARLY INTERNET

Bioinformatics began with the creation of individual tools and evolved into the provision of complete packages, which could be used to analyze DNA and protein sequences. Several of the original stand-alone tools were so essential early on in bioinformatics that they are still often used as individual tools today. We cannot make any attempt to list all of them, but some, which were important for the creation of genome analysis pipelines, are discussed next.

The FASTA and BLAST tools really opened up the ability to efficiently search the databases to almost anyone, as these tools allowed rapid searches on rather benign hardware and gave reasonable approximate answers within a short period of time.

Two multiple alignment tools really shine, and both were created by Des Higgins: Clustal (in its various incarnations up to the most recent ClustalX) (Thompson et al., 1994) and T-Coffee (Notredame et al., 1998). The original publication describing Clustal remains one of the most cited bioinformatics publications of all time and both tools are still in use in many laboratories. Figure 5.1 shows an example of using Clustal to perform multiple sequence alignment.

One problem, which began to haunt bioresearchers early on, was the large number of different file formats, which were developed without coordination

FIGURE 5.1 Command line screenshots of ClustalW performing multiple sequence alignment. A multiple sequence FASTA format file is entered (top), which CluatalW processes to produce an alignment format file containing the results of the multiple sequence alignment (bottom, initial part shown).

to store sequence information. Some of these formats, especially the FASTA and multiple FASTA file formats, are popular to this day. Don Gilbert from Indiana University solved this problem through the creation of a tool called Readseq (Gilbert, 2003), which can be used to convert many different file formats among each other. Readseq was one of the first public domain and open source bioinformatics tools and Gilbert can be credited as one of the early bioinformatics leaders who introduced this policy in the field.

Initially, users could only use local installations of bioinformatics tools, but even before the existence of the World Wide Web, developers began to make their tools accessible over the Internet. E-mail was used as a vehicle for the use of remote bioinformatics tools. The FASTA and BLAST e-mail servers, as well as some more "exotic" database search e-mail servers were commonly used. The major advantage that they provided was accessibility from almost any computer, including PC-type machines and the provision of up-to-date databases, unlike many local installations of the same search tools. Specialist tools, such as the BLOCKS server (Hennikoff and Hennikoff, 1991) emerged, which allowed the identification of protein motifs, and they were coming online. Still, e-mail servers could be easily overloaded, especially when the first automated tools began "talking" to them and they were ultimately replaced by Web interfaces.

One last toolkit, which is well worth mentioning, is the PHYLIP package (Felsenstein, 1989), which was created and is still being maintained by Joe Felsenstein and his colleagues. The ideas, which formed the basis for the creation of PHYLIP, are becoming more important in the age of environmental genomics, as the binning of sequences into operational taxonomic units (OTUs) is an essential task in this research field.

5.3 PACKAGES

It became clear early on in bioinformatics that a large gap existed between the tool developers and the end users. The installation and maintenance of programs proved to be very difficult for many and thus software packages emerged, which tried to address this gap through a unified software environment. Although today the computational platform for bioinformatics is usually a UNIX-based system, many different operating systems were initially explored in this context, as can be seen from the following descriptions of some of the more prominent early bioinformatics packages.

5.3.1 IBI/Pustell

One of the early software packages for sequence analysis was IBI/Pustell, written by John Pustell at Harvard University (Pustell and Kafatos, 1984). It was essentially created to handle manual sequencing and database searches. The package ran on MS-DOS computers and could be bought from IRL Press Software. As IRL Press was also distributing copies of the GenBank database at the time, the package could be used to manually enter data from autoradiograms and perform similarity searches against the included databanks.

5.3.2 PC/GENE

Users without access to a VMS or UNIX computer also demanded access to bioinformatics tools. Quite a number of software packages emerged early, from IBI-Pustell to what could probably be called one of the most useful MS-DOS-based suite of tools, PC/GENE (Moore et al., 1988), which was initially created by Amos Bairoch. The range of applications within PC/

GENE was similar to that of GCG/EGCG, but the package had the advantage that it could be used on any personal computer running MS-DOS and thus was also capable of printing on the printers connected directly to these machines. This was especially important for the graphics that PC/GENE created. Databank updates were available on floppy disks, which could be subscribed to, similar to the software updates for the program. Similar to GCG, memory restrictions made it difficult to impossible to handle large sequences. Workflows were not supported, thus users needed to store an output from one program and use it as input in another one subsequently.

5.3.3 GCG

As computers were initially not connected to any networks, they were considered quite secure and in many cases customers purchased a software package such as GCG, which was initially written for VAX-VMS and later recompiled for UNIX, together with the source code, which was then freely accessible for modifications. Within GCG, many of the early stand-alone bioinformatics applications, such as FASTA, BLAST, and others, were bundled. The system was command-line driven and the graphical outputs consisted of pixel graphics, mostly as black-and-white line drawings. Early on, database updates for the databank searches were performed from tapes, which were physically mailed by the databank providers. Later, as the Internet emerged, this was replaced by updates, which could be downloaded via anonymous FTP. In the case of GCG, the open software architecture, which allowed access to the source code, led to many additions to the toolkit, the most prominent being the EGCG package, which was created by Peter Rice (Fuchs et al., 1990). All executables within the package could only be used through a command-line interface (on monochrome text monitors). Over time, this became limiting and the GCG team added Steve Smith, the creator of an early bioinformatics user interface named GDE (Smith et al., 1994), to their roster and began using an adapted version of GDE as the interface to the GCG toolkit. The creation of workflows was only possible for experts with programming experience and certainly not a simple task for the average user of GCG.

5.3.4 From EGCG to EMBOSS

As said before, early versions of the GCG package included the source code, which allowed users to modify the code and recompile the GCG package. This was necessary in order to adapt the package to certain machine specifications, such as the available memory footprint. As these were the early days of what was later dubbed "bioinformatics," users also wanted features that were not included in the GCG distribution. Therefore some of the larger organizations, which had copies of GCG, began to add software components to the core distribution. Especially Peter Rice at EMBL in Heidelberg (this division has since relocated to Cambridge, England) was developing tools for GCG (Fuchs et al., 1990), which he distributed as the so-called EGCG (enhanced GCG) package. For a while, this was not only tolerated by the company owning the rights to GCG but even encouraged. Unfortunately the company distributing GCG eventually took a different route for the packaging of GCG, initially stopping the distribution of the source code and subsequently actively discouraging others from contributing to the code.

At this point in time, Rice and his team began a new, open source development, which led to the creation of the EMBOSS package (Mullan and Bleasby, 2002). This package, which has now essentially replaced GCG, is being adopted and supported by a large community of contributors, and therefore is now the *de facto* standard for a stand-alone, command-line driven bioinformatics package. EMBOSS is truly open source, therefore many contributors have added components to this toolkit over time. The use of EMBOSS is highly recommended for anyone looking for a solution to their basic bioinformatics needs. Several graphical and Web interfaces exist, which facilitate easy access to the EMBOSS tools.

5.3.5 The Staden Package

Eventually, automated DNA sequencing with fluorescent dyes, which are detected by photo cells and transformed automatically into "peaks," representing the base pairs of the DNA sequence, triggered the development of tools that could be used to handle this kind of information in a semiautomated fashion. Automated DNA sequencing naturally also led to an exponential increase in the number of sequences that could be generated, thus the completely manual, yet efficient handling of DNA sequences became more or less impossible.

One of the most prominent software packages that was used in this context was the Staden Package (Staden, 1986), which was created by Roger Staden and his team and maintained for many years. Roger Staden was the driving force behind the creation of trace data file formats. He created the original ABI trace data format (Staden, 1980) and later the SCF, Staden Chromatogram Format (Dear and Staden, 1992). The main purpose of the Staden Package was the extraction of sequence information from the trace

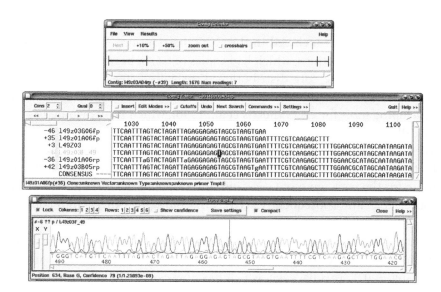

FIGURE 5.2 Viewing and editing contigs with the Staden Package. After sequence assembly from multiple trace files by the Gap4 program, contigs are shown in Contig Selector (top). Upon selecting a contig (the leftmost one in this case), Contig Editor shows the contig information including the reads used in constructing the consensus (middle), which can be manually edited. The trace around a base of a particular read can be viewed to aid in the editing process (bottom).

file, the trimming of the files to extract the useful information (after the removal of vector sequences, contaminating host DNA sequence, etc.) and the assembly of this information into contiguous sequences (i.e., contigs). In addition, the package provided many other tools, which were developed over time to assist with special tasks. Figure 5.2 shows an example of running Staden programs to build contigs from multiple trace files.

The Staden Package was built in a modular fashion, therefore other tools could be added to the data processing pipeline. Notable here are the Phred and Phrap tools (Ewing and Green, 1998; Ewing et al., 1998), which were created by Phil Green at the University of Washington. Phred allowed the creation of quality assessments for trace data and became more or less the *de facto* standard in this field until next-generation automated DNA sequencing took over.

The major tools in the Staden Package (pregap, the tool to manipulate trace data; and gap, the genome assembly program) gained graphical interfaces over time. This was essential, as the sequence assembly could be linked back to the trace data to verify sequence information and resolve any ambiguities through visual inspection.

5.3.6 GeneSkipper

The early sequencing efforts needed tools that were capable of analyzing larger contiguous sequence files (contigs) and providing additional information and graphical representations, which reflected the sequencing progress. EMBL Heidelberg was one of the sites involved in the *Saccharomyces cerevisiae* Yeast Genome Project and therefore Christian Schwager and his team at EMBL created a Windows-based software tool called GeneSkipper, which was one of the first tools to graphically represent an analysis of sequenced contigs (i.e., the six protein translation frames and potentially coding regions).

5.3.7 Sequencher

Apple Macintosh computers have always been considered to provide one of the most user-friendly user interfaces in the computer industry. Eventually, Howard Cash founded a company called GeneCodes, which created a sequence assembly system for the MacIntosh environment, called Sequencher (www.sequencher.com). Over time, Windows platform support has been added. This system is still being sold to this date and in use in many laboratories around the world.

5.4 FROM FASTA FILES TO ANNOTATED GENOMES

The Staden Package, GeneSkipper, and Sequencher all included very limited provisions for the genome annotation task. Essentially, they took raw sequence files in, and produced assembled sequence in FASTA format (or some other format) as the output. Therefore, dedicated tools were developed over time, initially to collect the annotation information, which was being created in a manual or semiautomated fashion (such as the ACeDB family of tools) and later on fully automated tools.

5.4.1 ACeDB

The *Arabidopisis thaliana* and *Caenorhabditis elegans* sequencing projects yielded AAtDB (an *Arabidopsis thaliana* database) (Cherry et al., 1992) and ACeDB (a *Caenorhabditis elegans* database) (Cherry and Cartinhour, 1994), which collected the information related to the genome annotation and graphically represented this information. Like GeneSkipper, these systems were created at the dawn of widespread Internet access. The genome annotations created with the ACeDB system could be downloaded and installed on UNIX machines (others might have been supported early on as well). The maintenance of this system was complicated and therefore creating a database for other genomes was undertaken only in a few cases, such as some human chromosomes and *Schizosaccharomyces pombe*. Especially, ACeDB handled completed genomic information best, as there was no notion of changing sequence states (from the first assembly to the finished genome) built into the workflow. ACeDB was retrofitted with Web capability later, but the success of this development was rather limited. Still, the systems have a Web presence to this day and a collection of these is currently maintained at the Sanger Center in the United Kingdom (www.acedb.org).

5.4.2 One Genome Project, the Beginning of Three Genome Annotation Systems

In the early days of complete genome sequencing, researchers usually kept the DNA sequence under wraps until it had been published. This, of course, prevented the community at large from access to these sequences and hence there was very little raw sequence information to go around, which could be used as an example while building bioinformatics tools that could be used to analyze and annotate large stretches of genomic sequence. There were a few notable differences, including the *Myoplasma capricolum* sequencing project (Dolan et al., 1995). The investigators opened access to early versions of the sequence in draft format. This eventually led to the creation of at least three software packages that played a role in the automated genome analysis and annotation field.

To this day, two of the systems, PEDANT (Frishman et al., 2001) and MAGPIE (Gaasterland and Sensen, 1996), are still in use, while the third, which was a commercial product called Bioscout, disappeared with the demise of LION Biosciences. PEDANT is now part of the offerings of Biomax Informatik AG in Munich, Germany, while MAGPIE is an open source system, which can be downloaded freely from Sourceforge. All three systems initially produced tabular HTML-based output of the functional assignments to genomic regions. Over time, the systems evolved and also added graphical representations of the analysis results. Some of the latest versions of the graphics-based tools will be introduced in more detail in subsequent chapters.

5.5 CONCLUSION

Bioinformatics as a research field is less than 40 years old. Within this time-frame, many useful stand-alone tools were developed, which can be used to characterize sequence information. Over time, they were combined into analysis pipelines, which can provide automated genome analysis and annotation for sequences. Initially, the annotations were listed in tabular format, but over time it became evident that graphical representations were more useful for most users of the annotation pipelines. Bioinformatics has truly adopted the Internet as the vehicle that is used to share the results of genome annotations. Almost every technological approach, from the early HTML-based annotation engines to the latest XML-based systems, has been tried over time. Today, we have efficient tools for the analysis and annotation of even the largest genomes, and systems for comparative approaches of even hundreds of thousands of genomes the size of the human genome are under development, as discussed in subsequent chapters of this book.

REFERENCES

Altschul, S.F., Gish, W., Miller, W., Myers, E.W., Lipman, D.J. 1990. Basic local alignment search tool. *J. Mol. Biol.* 5:403–410.

Benson, D., Lipman, D.J., Ostell, J. 1993. GenBank. *Nucleic Acids Res.* 1:2963–2965.

Bigelow, C.C. 1967. On the average hydrophobicity of proteins and the relation between it and protein structure. *J. Theor. Biol.* 16:187–211.

Cherry, J.M., Cartinhour, S.W. 1994. ACEDB, A tool for biological information. In *Automated DNA Sequencing and Analysis*, eds. M. Adams, C. Fields, C. Venter, 347–356. San Diego, CA: Academic Press.

Cherry, J.M., Cartinhour, S.W., Goodman, H.M. 1992. AAtDB, an *Arabidopsis thaliana* database. *Plant Mol. Biol. Rep.* 10:308–309.

Dear, S., Staden, R. 1992. A standard file format for data from DNA sequencing instruments. *DNA Seq.* 3:107–110.

Devereux, J., Haeberli, P., Smithies, O. 1984. A comprehensive set of sequence analysis programs for the VAX. *Nucleic Acids Res.* 12:387–395.

Dolan, M., Ally, A., Purzycki, M.S., Gilbert, W., Gillevet, P.M. 1995. Large-scale genomic sequencing: Optimization of genomic chemical sequencing reactions. *Biotechniques* 19:264–268, 270–274.

Doolittle, R.F. 1981. Similar amino acid sequences: chance or common ancestry? *Science* 214:149–159.

Edman, P. 1949. A method for the determination of amino acid sequence in peptides. *Arch. Biochem.* 22:475.

Ewing, B., Green, P. 1998. Base-calling of automated sequencer traces using phred. II. Error probabilities. *Genome Res.* 8:186–194.

Ewing, G., Hillier, L., Wendl, M.C., Green, P. 1998. Base-calling of automated sequencer traces using phred. I. Accuracy assessment. *Genome Res.* 8:175–185.

Felsenstein, J. 1989. PHYLIP—Phylogeny Inference Package (Version 3.2). *Cladistics* 5:164–166.

Frishman, D., Albermann, K., Hani, J., et al. 2001. Functional and structural genomics using PEDANT. *Bioinformatics* 17:44–57.

Fuchs, R., Stoehr, P., Rice, P., Omond, R., Cameron, G. 1990. New services of the EMBL Data Library. *Nucleic Acids Res.* 18:4319–4323.

Gaasterland, T., Sensen, C.W. 1996. Fully automated genome analysis that reflects user needs and preferences. A detailed introduction to the MAGPIE system architecture. *Biochimie* 78:302–310.

Gilbert, D. 2003. Sequence file format conversion with command-line readseq. *Curr. Protoc. Bioinformatics* Appendix 1:Appendix 1E.

Gilbert, W., Maxam, A. 1973. The nucleotide sequence of the lac operator. *Proc. Natl. Acad. Sci. USA* 70:1–4.

Hennikoff, S., Hennikoff, J.G. 1991. Automated assembly of protein blocks for database searching. *Nucleic Acids Res.* 19(23):6565–6572.

Kneale, G.G., Kennard, O. 1984. The EMBL nucleotide sequence data library. *Biochem. Soc. Trans.* 12:1011–1014.

Kolakowski, L.F., Leunissen, J.A., Smith, J.E. 1992. Prosearch: Fast searching of protein sequences with regular expression patterns related to protein structure and function. *Biotechniques* 13:919–921.

Lipman, D.J., Wilbur, W.J., Smith, T.F., Waterman, M.S. 1984. On the statistical significance of nucleic acid similarities. *Nucleic Acids Res.* 11:215–226.

Moore, J., Engelberg, A., Bairoch, A. 1988. Using PC/GENE for protein and nucleic acid analysis. *Biotechniques* 6:566–572.

Mullan, L.J., Bleasby, A.J. 2002. Short EMBOSS User Guide. European Molecular Biology Open Software Suite. *Brief. Bioinform.* 3:92–94.

Notredame, C., Holm, L., Higgins, D.G. 1998. COFFEE: An objective function for multiple sequence alignments. *Bioinformatics* 14(5):407–422.

Pearson, W.R., Lipman, D.J. 1988. Improved tools for biological sequence comparison. *Proc. Natl. Acad. Sci. USA* 85:2444–2448.

Pustell, J., Kafatos, F.C. 1984. A convenient and adaptable microcomputer environment for DNA and protein sequence manipulation and analysis. *Nucleic Acids Res.* 10:479–488.

Roberts, R.J., Macelis, D. 1993. REBASE—Restriction enzymes and methylases. *Nucleic Acids Res.* 21:3125–3127.

Sanger, F., Donelson, J.E., Coulson, A.R., Kössel, H., Fischer, D. 1973. Use of DNA polymerase I primed by a synthetic oligonucleotide to determine a nucleotide sequence in phage fl DNA. *Proc. Natl. Acad. Sci. USA* 70:1209–1213.

Smith, S.W., Overbeek, R., Woese, C.R., Gilbert, W., Gillevet, P.M. 1994. The genetic data environment: An expandable GUI for multiple sequence analysis. *Comput. Appl. Biosci.* 10:671–675.

Staden, R. 1980. A new computer method for the storage and manipulation of DNA gel reading data. *Nucleic Acids Res.* 25:3673–3694.

Staden, R. 1986. The current status and portability of our sequence handling software. *Nucleic Acids Res.* 10:217–231.

Thompson, J.D., Higgins, D.G., Gibson, T.J. 1994. Clustal W: Improving the sensitivity of progressive multiple sequences alignment through sequence weighting, position-specific gap penalties and weight-matrix choice. *Nucleic Acids Res.* 11:4673–4680.

CHAPTER 6

Visualization Techniques and Tools for Genomic Data

6.1 INTRODUCTION

Starting with the sequencing step, there are several stages involved in genome annotation including genome assembly, sequence alignment, phylogenetic analysis, and gene expression profiling, where the researcher might want to acquire a holistic, as well as detailed understanding of the genomic data that is currently being processed. Since human cognition capacity of any data presented in a textual format is relatively small, each stage in this sequence-to-annotation pipeline will undoubtedly benefit from the use of suitable visualization, which allows one to quickly perceive a wealth of information, which is presented in a limited space (such as a computer screen). This is particularly true when one needs to compare and contrast multiple related sets or streams of data in a unifying context to gain valuable insight into the nature of the genomic data, a commonly required task at all stages of genome annotation.

Prior to discussing individual visualization techniques and tools available, we would like to note two overarching characteristics of genomic data that are highly relevant to their visualization. First, huge quantities of data need to be dealt with, as visualization tools are often deployed in large-scale genomic analysis. For example, high-throughput DNA sequencing produces large amounts of sequence data in a relatively short time. We also need to access and utilize ever-increasing quantities of existing public data accessible from online data repositories, ranging from the DNA sequence files themselves to genome annotations, including taxonomic classifications. As a rule, some sort of semantic zooming mechanism (Loraine and Helt, 2002) is required to enable the user to explore the genomic data set in its entirety, as well as in specific detail. This in turn requires an underlying multiscale, or scalable representation of a complex data set, to accommodate the appropriate display at several different levels of detail.

Second, integration of heterogeneous types of genomic data is required, including (but not limited to) sequences, gene expression data, hierarchical ontologies and taxonomies, and textual annotation. This requirement directly influences several key design choices, with respect to data handling (loading and unloading), layout of visual display elements, and data normalization for meaningful integration. In many cases, new visual metaphors and associated graphical representations are developed to visualize new types of data or to facilitate new kinds of analyses. However, due to the lack of standardized visual representations for these different data types (O'Donoghue et al., 2010), it is not a trivial task for users to learn when and how to use a new integrative visualization tool.

6.2 VISUALIZATION OF SEQUENCING DATA

After base calling and quality value calculations by a sequencing machine, automated sequencing data processing begins. In *de novo* genome sequencing, sequence assembly is performed. In the case of resequencing projects, the reads need to be mapped to a reference genome. In either case, visual inspection of how the reads are aligned and assembled into larger structures (e.g., contigs and scaffolds) is beneficial in the interpretation and validation of the outputs, which were generated by automated read alignment and assembly programs. Recent advances in sequencing technology have resulted in unprecedented volumes of sequence reads, especially when dealing with the large numbers of short reads, which are now possible at a relatively low cost. With the emergence of these types of large-scale sequencing data, visualization plays a more important role than ever before. Existing alignment and assembly viewer programs have been expanded, and completely new visualization tools (e.g., for next-generation sequencing [NGS] data) have been developed to cope with the computational challenges that this has posed (Nielsen et al., 2010).

In the case of sequence assemblies, the basic function of read alignment and assembly viewers is to show the arrangement of the assembled sequence reads themselves, their alignment, and the resulting consensus sequence. In most of these viewers, sequence reads are shown as horizontally oriented strings of letters, which are stacked vertically to show their alignment. Thus the bases that constitute a column would contribute to the resulting consensus base, and strings of these consensus bases are shown at the top or bottom of the rows of sequence reads. Sequence read viewers generally offer several display features to aid the user in analyzing the alignment results, such as hiding or ghosting the bases that agree in a column to reduce visual clutter, highlighting the bases that are disagreeing with the consensus base using a different color, and representing quality values of the bases using gray scales.

The option of viewing original trace data used for base calling is usually available in most tools developed before NGS. For example, in popular read alignment and assembly viewers Gap4 (Bonfield et al., 1995) and Consed (Gordon et al., 1998), the trace viewing window can be launched from the usual read alignment view by clicking on the bases. Synchronization of cursor movement between the base view and the guiding line in the trace view makes it easy to see how ambiguity in the primary data causes uncertainty in base calling and consensus bases. See Chapter 5 (Figure 5.2) for a screenshot from the Gap4 program showing a read alignment view along with the trace data view window.

With the emergence of NGS, the way primary DNA sequencing data is analyzed has changed considerably. In the case of sequencing data generated by the Illumina Solexa Genome Analyzer (www.illumina.com) and the Applied Biosystems SOLiD system (www.appliedbiosystems.com), there are only image data saved and thus no trace data exist. Traditional read alignment viewers cannot be used to display these images, partly due to storage and display speed considerations. Moreover, inspection of a single read trace is less significant for NGS, due to the typical highly redundant sequence coverage of NGS experiments, which means that a base in question can be compared to a large number of other bases at the same position in all of the aligned sequence reads. Some alignment viewers also offer editing capability, which allows a user to change bases that influence the consensus. In addition, the breaking or joining of contigs is supported. These tools are also known as "finishing" programs, the most widely used ones being Gap4 (http://staden.sourceforge.net) (Bonfield et al., 1995) and Consed (www.phrap.org/consed/consed.html) (Gordon et al., 1998),

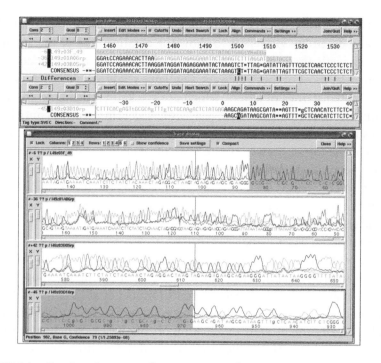

FIGURE 6.1   Gap4 environment for contig join editing. Two contigs produced by Gap4 are compared for possible joining, while reconsidering originally discarded lower-quality bases from all reads (in gray), with mismatching bases shown by exclamation marks in the differences row (top). The traces corresponding to the reads contributing to the currently examined portions of the contigs are also shown to aid in finding the potential join positions (bottom). **(See color insert.)**

both of which are freely available. Figure 6.1 shows an example of using Gap4 to join two assembled contigs. Commercial software suites such as Sequencher (www.genecodes.com) and Lasergene (www.dnastar.com) are also available.

As new sequencing technologies provide increased sequencing throughput at a lower cost than before, existing read alignment and assembly viewers were improved to cope with the challenge of higher volumes of NGS reads. Consed has been expanded and Gap5 (Bonfield and Whitwham, 2010) has been developed to be able to deal with substantially larger data sets. On the other hand, several new visualization tools are also being developed specifically for the purpose of handling NGS data.

EagleView (Huang and Marth, 2008) is a genome assembler and viewer that can read and visualize multiple reads in the standard ACE format as

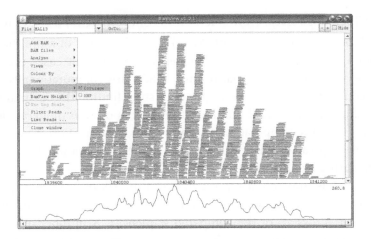

FIGURE 6.2   BamView display of read alignment of RNA-Seq data (obtained from   ftp://ftp.sanger.ac.uk/pub/pathogens/Plasmodium/falciparum/3D7/3D7. archive/misc/BAM/Plasmodium3D7_RNASeq.bam). The mapped reads are shown in a stack form with the corresponding coverage plot.

well as additional optional files, such as MAP files. It can also be used to visualize many types of information, including base qualities, traces, and genomic feature annotations. Navigation can be done by read or contig IDs, genomic features, or user-defined locations. These are the usual features that are also offered by the great majority of genome assembly viewers. MapView (Bao et al., 2009) can display alignments of large numbers of short reads and can also automatically detect genetic variation using a desktop computer. MapView uses a custom input file format (MVF, MapView Format) for the quick loading of NGS data. IGV (Robinson et al., 2011) is a genome browser with the ability to visualize NGS read alignments in the BAM (Binary Alignment/Map) format, a binary form of the SAM (Sequence Alignment/Map) format. SAMtools tview (Li et al., 2009) is a simple but fast text-based alignment viewer, where base qualities or mapping qualities are indicated by different colors. BamView (Carver et al., 2010) is an application to visualize large amounts of data stored for sequence reads, aligned to a reference genome and stored in BAM data files. Figure 6.2 shows an example of using BamView to visualize reads mapped to a reference genome. MagicViewer (Hou et al., 2010) is an alignment visualization and genetic variation detection/annotation tool for NGS data. Tablet (Milne et al., 2010) is an alignment and assembly viewer for NGS, with support for several input formats. It is built to be memory efficient, so that it can run even on a desktop computer.

6.3 VISUALIZATION OF MULTIPLE SEQUENCE ALIGNMENTS

A multiple sequence alignment (MSA) is an alignment of more than two biological sequences, where the sequences are in most cases presumed to have an evolutionary relationship. MSAs are computationally much more complex than pairwise alignments, since finding the optimal alignment of a relatively small number of sequences of even moderate length quickly becomes computationally hard. Thus optimization is performed by employing various heuristics to reduce the computational complexity of calculating an alignment.

An MSA is most often represented as a matrix of rows of sequences, in which each row is a sequence, and columns indicate positions that are equivalent across all of the compared sequences. A number of tools for visualizing MSAs have been developed, most of which use some variation on the basic display scheme of matrices of letters, along with some visual elements (e.g., coloring or shading) to highlight aligned nucleotides or amino acids within the overall alignment (Procter et al., 2010). Different visualization techniques can be used, depending on the number of sequences being compared (i.e., two versus multiple). Some tools allow users to edit and annotate the alignment results.

6.3.1 Pairwise Alignment Viewers

When only two sequences are being aligned, two-dimensional plots can be used to represent their degrees and locations of similarity intuitively. Dot plots show the comparison of two sequences by placing one sequence along the horizontal axis and the other one along the vertical axis, and visualizing their relationships using dots placed on the two-dimensional space thus formed. When a base or residue of one sequence matches that of the other sequence, a dot is drawn at the corresponding point. In the idealized case of comparing two sequences that are identical over the two ranges plotted, a continuous diagonal line will be formed from one corner of the plot to the diagonally opposite corner. In general, a number of diagonal line segments (or isolated dots) of various lengths will be drawn, representing the locations of the matches. Dot plots can only be used for comparing exactly two sequences. An early description of the use of dot plots in comparing two sequences can be found in Maizel and Lenk (1981).

One limitation of dot plots is their inability to express the degree of similarity between two segments of sequences, unless an additional visual cue (e.g., color) is used to differentiate between different levels of similarity. Percent identity plots (PIPs) show one sequence on the horizontal axis, while

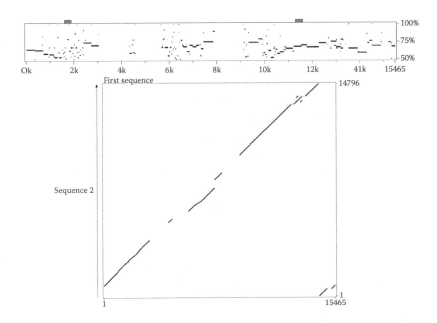

FIGURE 6.3   PipMaker comparison of two sequences. Using the first sequence as the reference (positions horizontally shown), the percent identity plot shows match with the second sequence as horizontal line segments, with their vertical positions indicating percent identities (top). The dot plot shows only the matched positions as dots, without sequence similarity information (bottom).

the vertical axis represents the percent identity. At horizontal positions of the first sequence where there is a significant match (e.g., greater than 60%) between the two sequences, horizontal line segments are drawn with their vertical positions corresponding to the percent identity of the two sequences in those positions. The locations and degrees of similarity between two sequences are readily shown. For example, PipMaker (Schwartz et al., 2000) is a tool with a Web interface, which can be used to compute alignments of two sequences that are supplied by a user. The tool generates PIPs, as well as textual forms of the alignment. Figure 6.3 shows an example of a PIP along with the corresponding dot plot, both generated by PipMaker. PIPs can also be used for comparing multiple sequences, by treating one of them as a reference sequence and the rest as query sequences, respectively. The comparison results in this case will be a collection of PIPs, where each one corresponds to a single comparison of a query sequence with the reference sequence. Another sequence alignment tool offering PIP is MUMmer 3 (http://mummer.sourceforge.net), which also generates dot plots.

6.3.2 Multiple Alignment Viewers

Visualization of multiple alignment results usually takes the form of a matrix of letters representing bases or residues. Two widely used multiple alignment programs are ClustalW (Thompson et al., 2002) and T-Coffee (Notredame et al., 2000), both of which were originally text based. ClustalW also offers a user-friendly graphical version called ClustalX (Thompson et al., 2002). This type of graphical interface allows a user to adjust visualization elements such as color and shading to produce different visual representation of alignment, and also figures for use in publications. ClustalW is currently also available through the Web servers at European Bioinformatics Institute (www.ebi.ac.uk/Tools/clustalw2) and at Swiss Institute of Bioinformatics (www.ch.embnet.org/software/ClustalW.html). T-Coffee and its variants, such as M-Coffee (Wallace et al., 2006) for combining results from popular alignment tools, can be accessed at the T-Coffee Web site (www.tcoffee.org) and its mirror sites.

Coloration is frequently used in multiple alignment visualization to highlight the dominance of specific properties in a region and to indicate sequence variation at the same time. The most frequently used coloring method is to adopt a fixed color-mapping scheme, where each nucleotide or amino acid is assigned a unique color. There is no universal coloring standard, and each viewer defines its own coloring scheme, with some allowing user-defined schemata as well.

Most multiple alignment viewers generate some type of annotation. The most common one is the consensus row at the top or bottom of the alignment matrix, which is used to summarize the sequence alignment results. The calculated alignment score at each column can also be represented in a separate line as a bar plot, where the bar heights correspond to the alignment qualities of the columns. For example, Figure 6.4 shows a screenshot of the Jalview tool (Waterhouse et al., 2009), showing alignment of multiple sequences along with such a bar plot as well as a consensus plot. These plots are collectively known as summary plots and can contain additional information, especially the dominant symbol at each aligned position. A set of symbols can also be shown at each position, where the relative sizes of the symbols would indicate the dominance of the pattern.

Multiple alignment algorithms are designed to allow users to find the best solution for a biological problem that has been mathematically formulated. As such, even the most sophisticated alignment algorithms can generate alignments that are not biologically meaningful. In addition,

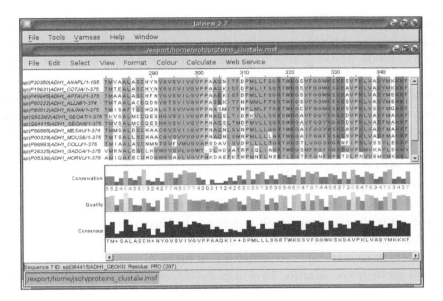

FIGURE 6.4   Jalview showing multiple sequence alignment of related proteins. Various coloring schemes can be used to help visualize the alignment (Clustalx selected in the example), with summary plots for alignment quality and consensus (bottom). **(See color insert.)**

multiple alignment algorithms rely on heuristics to make finding the best alignment computationally feasible, and they do not necessarily produce the perfect alignment. This means that their results often contain errors. Detecting and correcting these errors requires special knowledge, thus the task is often difficult to automate. Therefore, it is often necessary to manually edit and curate automatically generated alignments. Many alignment visualization tools also offer this editing capability with sequence navigation and search support to help the user quickly locate the region of interest for detailed analysis and subsequent editing of the alignment. Representative tools that allow editing and curation include GeneDoc (www.psc.edu/biomed/genedoc) (Nicholas et al., 1997), Jalview (Waterhouse et al., 2009), PFAAT (Caffrey et al., 2007), and CINEMA (Lord et al., 2002).

6.4 VISUALIZATION OF HIERARCHICAL STRUCTURES

Many concepts or findings in biological studies are best described as a hierarchy, often represented in the form of a tree. For example, phylogenetic analysis generates phylogenetic trees, which are treelike representations of inferred evolutionary relationships among the respective organisms being studied. Another example is the Gene Ontology project (Ashburner et al., 2000), whose goal is to provide a controlled vocabulary of terms for describing instance gene and gene product attributes and annotation data across all species. Biological ontologies may contain a large number of terms that are best organized into hierarchical structures represented as trees or graphs. Clustering is an analysis method applied to a wide variety of biological studies, such as microarray experiments, sequence analysis, and sometimes phylogenetic analysis, with the goal of grouping and partitioning data elements based on their distances or similarities. Some commonly used clustering algorithms, for example, hierarchical clustering, frequently show the relationships among the clusters as a tree, as the clusters are generated and joined to form bigger clusters. As hierarchical tree structures are natural representations in many biological analyses, many tools for their visualization have been developed (Pavlopoulos et al., 2010).

6.4.1 Tree Visualization Styles

Most visualization tools have built-in functions to display a tree using Euclidean geometry, often using several different styles, especially when displaying phylogenetic trees. A phylogenetic tree can be visualized as either a phylogram or a cladogram. In a phylogram, branch lengths are thought to be in proportion to the degree of evolutionary change, whereas in a cladogram, the branching order represents the inferred phylogeny. In addition, each tree type may be rooted or unrooted, depending on the positioning of the species with respect to an out-group sequence. A phylogram or cladogram with a hypothetical common ancestor (i.e., the root) is therefore considered to be a rooted tree, and one without such an ancestor is considered an unrooted tree.

In terms of the spatial layout, the shape of a tree may be rectangular, slanted, or circular. In a rectangular tree, the nodes are laid out along the horizontal or vertical axis, and the branches are drawn along the other axis to show the hierarchical structure. Hundreds of nodes can be visualized in a rectangular tree, often in association with other rectangle-based displays, such as heat maps, but visual navigation by the user following the branches can quickly become difficult. A slanted tree has essentially the same layout as the rectangular version, except that it uses sloped lines to draw branches to make the hierarchy better stand out. A circular tree has its root at the center of a circular space, with branches radiating out in all directions from the center. This layout is more space efficient than the

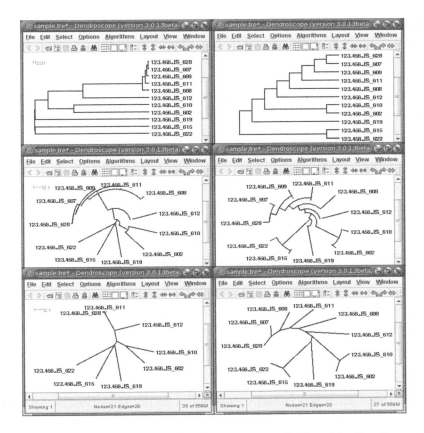

FIGURE 6.5   Different layouts of a clustering tree displayed in Dendroscope. Shown from left to right, and top to bottom are the rectangular phylogram, rectangular cladogram, circular phylogram, circular cladogram, radial phylogram, and radial cladogram of the tree.

rectangular or slanted layout, but it may be more challenging to compare branch lengths. Figure 6.5 illustrates six different layouts of the same clustering tree, as generated using Dendroscope (Huson et al., 2007).

For a tree with a large number of levels and many children per node, the numbers of nodes and edges to be displayed grow exponentially, requiring an enormous amount of space and causing severe visual clutter when any of the Euclidean layouts is used. A more efficient use of space can be achieved by drawing trees in a hyperbolic space. This layout is suitable for visualizing a very large number of nodes. Hyperbolic trees are usually displayed within a circle, which gives a fish-eye (i.e., wide-angle) lens view of the hyperbolic plane. The key idea is to emphasize the nodes that are in focus by positioning them toward the center of the circle, while the nodes that are out of focus are placed off the center toward the circle boundary. The edge lengths are scaled exponentially based on their distances from the circle center. By providing a panning mechanism, hyperbolic tree viewers allow the user to bring different branches of a tree into the central focused area for a closer examination, while at the same time maintaining the context of the whole tree.

6.4.2 Tree Visualization Tools

There is no official standard file format for describing a tree. However, there exists a widely used format supported by most tree visualization tools: the Newick format. Also known as the New Hampshire format, the Newick format uses character strings to name nodes, pairs of parentheses to represent the children of a node, and commas to separate nodes. A number can be attached to any node to indicate its distance from the parent node. A detailed description and examples can be found from the Web page maintained by the author of the PHYLIP package (Felsenstein, 1989), who was instrumental in popularizing the Newick format (http://evolution.genetics.washington.edu/phylip/ newicktree.html). The Nexus format is another commonly used format, which includes a Newick-formatted tree together with meta-data support. This format was initially introduced by the Phylogenetic Analysis Using Parsimony (PAUP) (Swofford, 2002), a widely used package for the inference of phylogenetic trees.

TreeView (Page, 1996) is a classic program for displaying phylogenies on Apple Macintoshes and Windows PCs. It provides a simple way to view the contents of a tree in common tree format files, with the ability to read trees with up to 1000 taxa. Supports are provided for native graphics file formats (i.e., PICT on Macintosh and Windows Metafile on Windows) to copy tree images into other applications and to save trees as image files. Printing multiple trees on a page or a single tree over multiple pages is possible. Trees can also be edited. TreeView is not under active development anymore, but it was one of the earliest programs to use for quick tree visualization. TreeView X (http://darwin.zoology.gla.ac.uk/~rpage/treeviewx) was developed later as a version of TreeView to support the display of phylogenetic trees on Linux and other Unix systems.

Dendroscope (Huson et al., 2007) can be used to visualize large rooted phylogenetic trees and networks, often involving hundreds of thousands of taxa. It allows the user to manipulate and edit trees interactively, with features such as magnification, collapsing and expanding of subtrees, label editing, and rerooting trees. Querying tree data using regular expressions is also possible. With multiple input trees, consensus trees and rooted phylogenetic networks can be derived. Several tree layout algorithms including rectangular, slanted, circular, and radial views are available. Dendroscope accepts the Newick and Nexus input formats, and displayed trees can be exported in one of several common image and document formats. Written in Java, the system is computing-platform independent.

Interactive Tree Of Life (iTOL) (Letunic and Bork, 2007) is a Web-based application for phylogenetic tree display and manipulation. iTOL supports three different layouts that allow users to display a phylogenetic tree: normal (rectangular), circular, and unrooted (radial). The system offers several options for the customization of tree visualization, such as pruning trees, collapsing clades, rerooting trees, coloring subtrees, and rotating branches. Trees in the Newick format or the Nexus format can be read, and trees can be exported to several graphics formats. iTOL is most suitable for visualizing midsized trees, such as trees with up to several thousand leaves. iTOL maintains several precomputed trees available for display, including the main Tree of Life (Ciccarelli et al., 2006), which is a tree representation of the evolutionary relationship of 191 species for which completely sequenced genomes are available.

Some tree viewers can handle very large phylogenetic trees by using visualization in a hyperbolic space. Hypergeny (http://bioinformatics.psb. ugent.be/hypergeny/home.php) is a Java application that can also be used through the Web for the visualization of large phylogenetic trees using a hyperbolic tree browser. It takes a Newick formatted tree file as input, and can be used to export trees and subtrees in the same format. Figure 6.6 shows an example of using Hypergeny that illustrates that a hyperbolic tree can have different foci, but still the whole tree is displayed within the application view. HyperTree (Bingham and Sudarsanam, 2000) is another Java application developed for viewing and manipulating large hierarchical data sets. It has many features to help the user navigate through large trees, such as searching and selecting nodes, coloring subtrees, labeling branches, and a zooming control.

Phylogenetic trees often have branch lengths that represent evolutionary distances. This display mode can be considered a basic type of annotation. Further annotation of phylogenetic trees with additional information from existing data sources can facilitate the better biological interpretation of those trees. TreeDyn (Chevenet et al., 2006) is a tree visualization

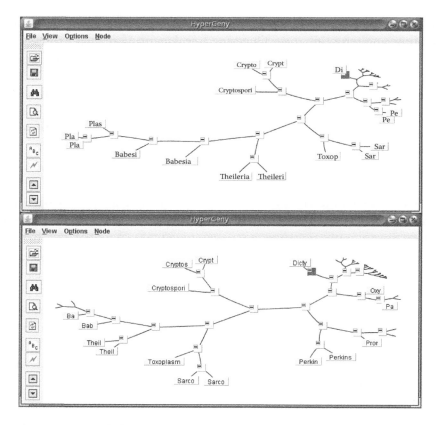

FIGURE 6.6   Two different views of a hyperbolic tree displayed with Hypergeny (a sample tree that comes with Hypergeny package). The top and bottom views have different foci and tree branch distances are scaled in a hyperbolic space following non-Euclidean geometry.

and annotation tool that can handle multiple trees and use meta-information to annotate trees through interactive graphical operators. iTOL v2 (Letunic and Bork, 2011) is an extension of the existing iTOL, with an emphasis on annotation capability.

6.5 VISUALIZATION OF GENE EXPRESSION DATA

High-throughput gene expression profiling methods, such as DNA microarrays, made it possible to acquire a snapshot of the activity of hundreds or even thousands of genes expressed in the organism. Visualization techniques have become indispensable for the analysis and interpretation of the enormous amounts of data produced by gene expression studies, irrespective of using gene chips or high-throughput DNA sequencing as the measurement method. The general goal of gene expression profile analysis is to find a set of genes that are overexpressed or underexpressed under some conditions, for example, in a disease model or in response to a prescribed treatment. The difficulty with accurate detection of these target genes mainly arises from overwhelming amounts of data to be analyzed simultaneously. It is not uncommon for a single data set to contain expression profiles of more than 10,000 genes. The same kind of data set could be generated under different experimental conditions or at certain time intervals over a period of time, thereby further increasing data set sizes. Clustering analysis and visualization are two key approaches to tackle this data interpretation problem. A review of gene expression data analysis and visualization can be found in Quackenbush (2001).

6.5.1 Expression Data Visualization Techniques

Gene expression studies usually produce multivariate data, which require efficient use of the two-dimensional space to visualize the results. There are many such visualization techniques, three of which are frequently used in gene expression data visualization: heat maps, profile plots, and scatter plots. Of these the generation of heat maps is the most commonly used gene expression visualization method mainly because it is relatively easy and straightforward to create from raw data. Clustering is usually the first analysis applied to most kinds of gene expression data in order to detect groups of genes that behave in similar ways under multiple experimental conditions. A comprehensive list of visualization tools for multivariate data, including gene expression data, is available from Gehlenborg et al. (2010).

Heat maps are created by first defining a color gradient to represent expression levels and subsequently drawing a small grid for each gene under the experimental condition or at a time point. Essentially a heat map is a matrix of these grids, where each row corresponds to a gene and each column corresponds to an experimental observation, be it a tissue, a disease condition, or a time point. Thus the size of the heat map increases quickly, proportional to the number of genes being observed at the same time. Rows of a heat map can usually be reordered to reflect the similarity between groups of genes, for example, to show genes clustering together as a result of hierarchical clustering. It is also simple to add annotations regarding gene names and experiments using the space around the matrix. Despite these advantages, heat maps are not really the most intuitive visualization method, since grid colors always need to be mentally mapped to the direction of regulation (i.e., up or down) and magnitude of expression levels for any interpretation of observations to occur, and there is no agreed-upon standard coloring scheme at this point in time.

Profile plots are a common way to visualize high-dimensional data in a two-dimensional space. To plot n-dimensional data points, n vertical and equally spaced lines are drawn. A point in an n-dimensional space is represented as a line graph, connecting n vertices on the n parallel lines, where the vertex positions correspond to the n coordinates of the point. Thus profiles plots are also known as parallel coordinate plots. This visualization method enables the user to perceive the expression trends of many genes at the same time, as the vertical positions on the parallel lines can easily be visually compared. This method facilitates the comparison of expression levels of a set of genes under different experimental conditions (e.g., normal versus diseased tissue) or over a course of time. It is also possible to quickly find a set of genes that are upregulated or downregulated at certain experimental conditions of time points. The biggest limitation of profile plots is that the line graphs tend to severely overlap, creating visual clutter, which may make it hard to visualize expression profiles of even a moderate number of genes at the same time. As a solution to this problem, coloring schemes can be adopted to help distinguish different graph lines. Figure 6.7 shows expression profile plots generated after clustering of genes in the TIGR MultiExperiment Viewer (MeV) (Saeed et al., 2003).

In a scatter plot, a collection of data points is plotted to show the relationship between two variables. If there is a clear correlation between the two variables, scatter plots can be used to show this correlation between them. In case of gene expression analysis, there are always more than two variables, requiring a reliable method of transforming the original data into two-dimensional data. As a result, a correct transformation would place two genes with similar expression profiles close to each other in a two-dimensional scatter plot. In most cases, dimensionality reduction needs to be performed to extract the two variables from the data. One popular method is principal component analysis (PCA) (Wold et al., 1987), which is a mathematical procedure to reduce multiple, possibly correlated variables into a smaller number of uncorrelated variables, which are called principal components, such that the observed data has the highest variation with respect to the first principal component. The subsequent components would account for less and less variation in the data. Two-dimensional scatter plots can only show data variation with respect to the first two principal components, whereas three-dimensional plots may be used to show

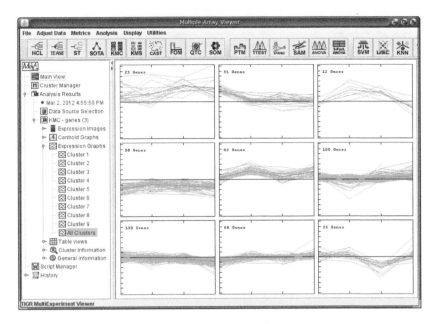

FIGURE 6.7 TIGR MeV display of nine expression profile plots, showing the result of clustering genes based on expression values over four different tissues. In each profile, the darkest gray line (excluding the midline) represents the mean profile of the genes in the cluster.

variation in the first three components. Scatter plots are good for obtaining a first insight into the general characteristics of the data, but the chief drawback is that the plotted data points cannot be mapped back to the observed data (e.g., to extract corresponding experimental conditions), as they have already gone through an irreversible dimensionality reduction procedure.

6.5.2 Visualization for Biological Interpretation

The visualization techniques described so far focus on gene expression data only, without showing any information on the biological meaning that the expression profiles might imply. A new direction in gene expression analysis is to link gene expression profiles to the functional annotation of genes. To achieve this, in essence expression profiles are mapped to functional profiles, such that gene functions that are upregulated or downregulated can be shown, in addition to, or instead of simple gene names. Most visualization tools that are capable of doing this mapping use the Gene Ontology (GO) to allow the classification of gene function, which offers standardized terminology for gene function annotation.

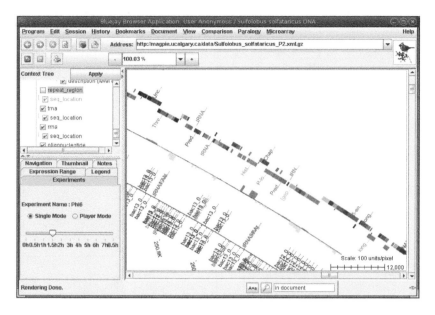

FIGURE 6.8   Bluejay display of gene expression data along a genome. The red and green bars along the inner circle represent upregulated and downregulated genes, respectively, with the bar heights indicating gene expression levels. The bars are aligned to the radial positions of the corresponding genes shown in the outer circle. **(See color insert.)**

Most traditional visualization techniques for gene expression data do not currently have at this point in time a provision of linking the observed data to the genomes of the organism from which the expression levels were measured. Visualization of gene expression levels along a genome is useful for providing the context of the whole genome in interpreting the expression values. For example, the Bluejay genome browser (Soh et al., 2012; Soh et al., 2008) can be used to display expression levels as bar graphs directly alongside the genome, by integrating the TIGR MeV gene expression viewer within the browser and adding a new display feature to the Bluejay system. Figure 6.8 shows a section of a genome with corresponding gene expression levels plotted as red and green bars, facilitating direct association of expression levels and the genes for which they were measured. Combined with the existing annotation visualization capability of the genome browser, this can help the user to gain more insight about the particular expression profiles being visualized such as an operon structure (see Section 4.2). This is a major advantage of the Bluejay system when compared to the usual display of expression values in isolation.

REFERENCES

Ashburner, M., Ball, C.A., Blake, J.A., et al. 2000. Gene ontology: Tool for the unification of biology. The Gene Ontology Consortium. *Nat. Genet.* 25:25–29.

Bao, H., Guo, H., Wang, J., Zhou, R., Lu, X., Shi, S. 2009. MapView: Visualization of short reads alignment on a desktop computer. *Bioinformatics* 25(12):1554–1555.

Bingham, J., Sudarsanam, S. 2000. Visualizing large hierarchical clusters in hyperbolic space. *Bioinformatics* 16:660–661.

Bonfield, J.K., Smith, K., Staden, R. 1995. A new DNA sequence assembly program. *Nucleic Acids Res.* 23:4992–4999.

Bonfield, J.K., Whitwham, A. 2010. Gap5—Editing the billion fragment sequence assembly. *Bioinformatics* 26:1699–1703.

Caffrey, D.R., Dana, P.H., Mathur, V., et al. 2007. PFAAT version 2.0: A tool for editing, annotating, and analyzing multiple sequence alignments. *BMC Bioinformatics* 8:381.

Carver, T., Bohme, U., Otto, T.D., Parkhill, J., Berriman, M. 2010. BamView: Viewing mapped read alignment data in the context of the reference sequence. *Bioinformatics* 26:676–677.

Chevenet, F., Brun, C., Banuls, A.L., Jacq, B., Christen, R. 2006. TreeDyn: Toward dynamic graphics and annotations for analyses of trees. *BMC Bioinformatics* 7:439.

Ciccarelli, F.D., Doerks, T., von Mering, C., Creevey, C.J., Snel, B., Bork, P. 2006. Toward automatic reconstruction of a highly resolved tree of life. *Science* 311:1283–1287.

Felsenstein, J. 1989. PHYLIP—Phylogeny Inference Package (Version 3.2). *Cladistics* 5:164–166.

Gehlenborg, N., O'Donoghue, S.I., Baliga, N.S., et al. 2010. Visualization of omics data for systems biology. *Nature Methods Supplement* 7:S56–S68.

Gordon, D., Abajian, C., Green, P. 1998. Consed: A graphical tool for sequence finishing. *Genome Res.* 8:195–202.

Hou, H., Zhao, F., Zhou, L. 2010. MagicViewer: Integrated solution for next-generation sequencing data visualization and genetic variation detection and annotation. *Nucleic Acids Res.* 38:W732–W736.

Huang, W., Marth, G. 2008. EagleView: A genome assembly viewer for next-generation sequencing technologies. *Genome Res.* 18:1538–1543.

Huson, D.H., Richter, D.C., Rausch, C., Dezulian, T., Franz, M., Rupp, R. 2007. Dendroscope: An interactive viewer for large phylogenetic trees. *BMC Bioinformatics* 8:460.

Letunic, I., Bork, P. 2007. Interactive Tree Of Life (iTOL): An online tool for phylogenetic tree display and annotation. *Bioinformatics* 23:127–128.

Letunic, I., Bork, P. 2011. Interactive Tree Of Life v2: Online annotation and display of phylogenetic trees made easy. *Nucleic Acids Res.* 39:W475–W478.

Li, H., Handsaker, B., Wysoker, A., et al. 2009. The Sequence Alignment/Map format and SAMtools. *Bioinformatics* 25:2078–2079.

Loraine, A.E., Helt, G.A. 2002. Visualizing the genome: techniques for presenting human genome data and annotations. *BMC Bioinformatics* 3:19.

Lord, P.W., Selley, J.N., Attwood, T.K. 2002. CINEMA-MX: A modular multiple alignment editor. *Bioinformatics* 18:1402–1403.

Maizel, J.V., Jr., Lenk, R.P. 1981. Enhanced graphic matrix analysis of nucleic acid and protein sequences. *Proc. Natl. Acad. Sci. USA* 78:7665–7669.

Milne, I., Bayer, M., Cardle, L., et al. 2010. Tablet—Next generation sequence assembly visualization. *Bioinformatics* 26:401–402.

Nicholas, K.B., Nicholas, H.B. Jr., Deerfield, D.W. II. 1997. GeneDoc: Analysis and visualization of genetic variation. *EMBNET.news* 4:1–4.

Nielsen, C.B., Cantor, M., Dubchak, I., Gordon, D., Wang, T. 2010. Visualizing genomes: Techniques and challenges. *Nat. Methods* 7(3 Suppl.):S5–S15.

Notredame, C., Higgins, D.G., Heringa, J. 2000. T-Coffee: A novel method for fast and accurate multiple sequence alignment. *J. Mol. Biol.* 302:205–217.

O'Donoghue, S.I., Gavin, A.C., Gehlenborg, N., et al. 2010. Visualizing biological data-now and in the future. *Nat. Methods* 7(3 Suppl.):S2–S4.

Page, R.D. 1996. TreeView: An application to display phylogenetic trees on personal computers. *Comput. Appl. Biosci.* 12:357–358.

Pavlopoulos, G.A., Soldatos, T.G., Barbosa-Silva, A., Schneider, R. 2010. A reference guide for tree analysis and visualization. *BioData Mining* 3:1.

Procter, J.B., Thompson, J., Letunic, I., Creevey, C., Jossinet, F., Barton, G.J. 2010. Visualization of multiple alignments, phylogenies and gene family evolution. *Nat. Methods* 7(3 Suppl.):S16–S25.

Quackenbush, J. 2001. Computational analysis of microarray data. *Nat. Rev. Genet.* 2:418–427.

Robinson, J.T., Thorvaldsdóttir, H., Winckler, W., et al. 2011. Integrative genomics viewer. *Nat. Biotechnol.* 29:24–26.

Saeed, A.I., Sharov, V., White, J., et al. 2003. TM4: A free, open-source system for microarray data management and analysis. *Biotechniques* 34(2):374–378.

Schwartz, S., Zhang, Z., Frazer, K.A., et al. 2000. PipMaker—A Web server for aligning two genomic DNA sequences. *Genome Res.* 10:577–586.

Soh, J., Gordon, P.M.K., Sensen, C.W. 2012. The Bluejay genome browser. *Curr. Protoc. Bioinformatics* 10.9.1–10.9.23.

Soh, J., Gordon, P.M.K., Taschuk, M.L., et al. 2008. Bluejay 1.0: Genome browsing and comparison with rich customization provision and dynamic resource linking. *BMC Bioinformatics* 9:450.

Swofford, D.L. 2002. *PAUP: Phylogenetic Analysis Using Parsimony (and Other Methods) 4.0 Beta.* Sunderland, MA: Sinauer Associates.

Thompson, J.D., Gibson, T.J., Higgins, D.G. 2002. Multiple sequence alignment using ClustalW and ClustalX. *Curr. Protoc. Bioinformatics* 2:2.3.1–2.3.22.

Wallace, I.M., O'Sullivan, O., Higgins, D.G., Notredame, C. 2006. M-Coffee: Combining multiple sequence alignment methods with T-Coffee. *Nucleic Acids Res.* 34:1692–1699.

Waterhouse, A.M., Procter, J.B., Martin, D.M., Clamp, M., Barton, G.J. 2009. Jalview Version 2—A multiple sequence alignment editor and analysis workbench. *Bioinformatics* 25:1189–1191.

Wold, S., Esbensen, K., Geladi, P. 1987. Principal component analysis. *Chemometr. Intell. Lab.* 2:37–52.

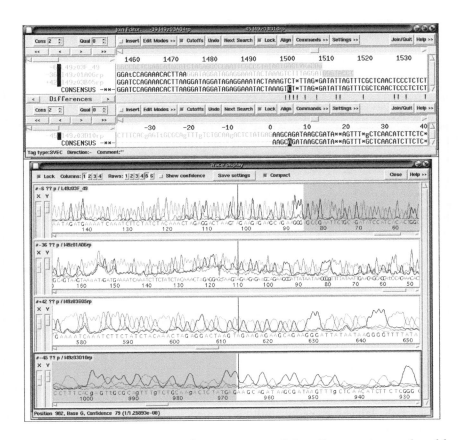

FIGURE 6.1   Gap4 environment for contig join editing. Two contigs produced by Gap4 are compared for possible joining, while reconsidering originally discarded lower-quality bases from all reads (in gray), with mismatching bases shown by exclamation marks in the differences row (top). The traces corresponding to the reads contributing to the currently examined portions of the contigs are also shown to aid in finding the potential join positions (bottom).

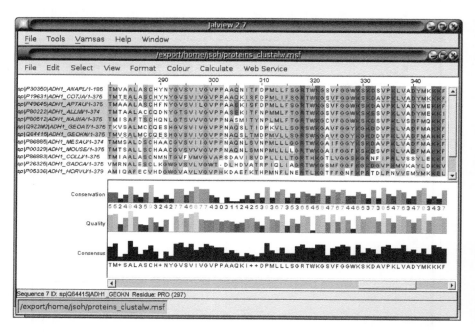

FIGURE 6.4 Jalview showing multiple sequence alignment of related proteins. Various coloring schemes can be used to help visualize the alignment (Clustalx selected in the example), with summary plots for alignment quality and consensus (bottom).

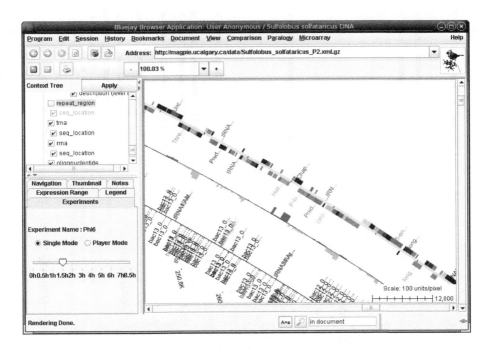

FIGURE 6.8 Bluejay display of gene expression data along a genome. The red and green bars along the inner circle represent upregulated and downregulated genes, respectively, with the bar heights indicating gene expression levels. The bars are aligned to the radial positions of the corresponding genes shown in the outer circle.

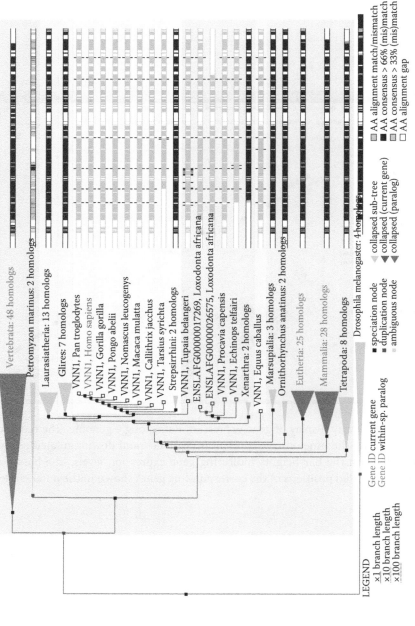

FIGURE 7.3 Visual gene tree depiction in Ensembl for the VNN1. Although most species have one copy of the orthologous gene, *Loxodonta africana* has two, showing that orthologs are not necessarily one-to-one.

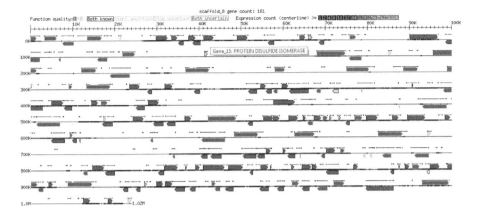

FIGURE 8.2 MAGPIE genomic sequence structural overview. Each arrow box represents a gene on either strand. The broken lines above these show the exon structure of each transcript, along with demarcations of the gene starts and stops (vertical cap lines). Mapped RNA-Seq read density is shown in red (in varying brightness) on the centerline. Element and outline coloration is explained in the legend at the top of the images.

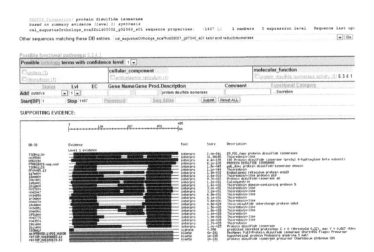

FIGURE 8.4 MAGPIE functional annotation form. From top to bottom: a textual description synthesized from all of the evidence, a list of possible paralogs/homologs, biochemical pathway information, Gene Ontology terms, an edit facility for the functional annotations, start/stop editing, and analysis evidence. Clicking the analysis evidence shows the original tool output, whereas clicking the ID leads to the public database entries.

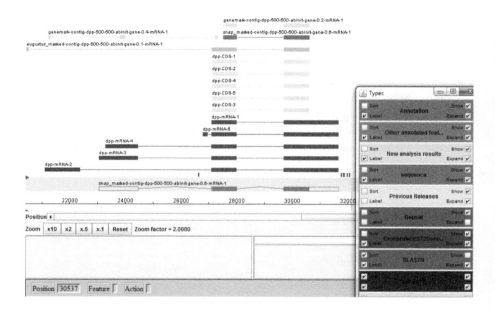

FIGURE 8.7   Apollo genome browser view of MAKER genomic structural annotation. A consensus gene structure is displayed at the bottom (light blue background color), with contributing evidence stacked on top (different colors for different evidence sources according to the legend at right). Apollo allows graphical editing of model borders, with write-out to GFF file format.

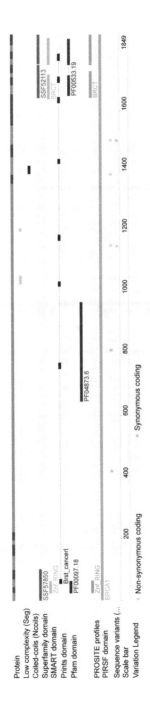

FIGURE 8.9 Ensembl functional evidence summary. (Top) Exon structure, with alternating light and dark purple to show boundaries. (Middle) Two biochemical feature and six domain homology lines of evidence. (Bottom) Known sequence variants (when available).

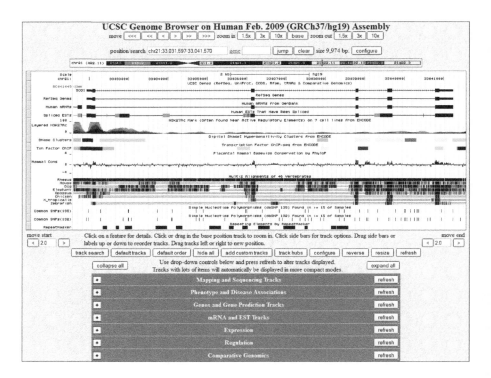

FIGURE 9.1 UCSC Genome Browser display of a human genome assembly. The top portion contains zoom scale and position control buttons, the middle portion is the main graphic showing various annotations, and the bottom portion has the menus for controlling display of tracks.

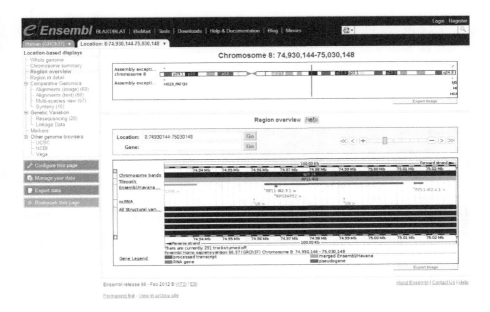

FIGURE 9.2 Ensembl Genome Browser display of Region Overview. The top portion (chromosome view) indicates the currently viewed region within the chromosome and the bottom portion shows the genomic features in several tracks. The bar in the middle allows the user to navigate to other regions or change the viewing scale.

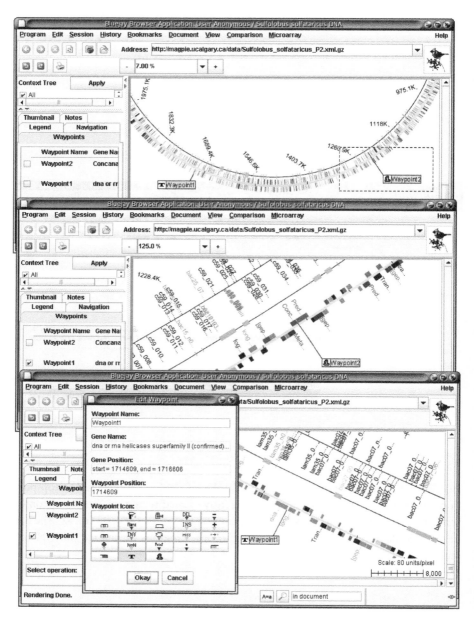

FIGURE 9.4   Customized navigation using waypoints in Bluejay. The user sets two waypoints on the displayed genome and zooms in on the area around "Waypoint2" (top). The selected region is zoomed in and displayed in more detail. Then the user clicks on "Waypoint1" in the "Waypoints" tab (middle). The focus changes to the region around "Waypoint1." Then the user selects "Edit Waypoint" operation to change the attributes of "Waypoint1" (bottom).

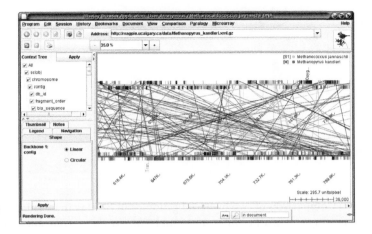

FIGURE 9.6   Comparison of two archaeal genomes in Bluejay in linear representation. Comparing *Methanococcus jannaschii* and *Methanopyrus kandleri* shows that they share many genes in the same Gene Ontology (GO) categories, indicated by the linking lines.

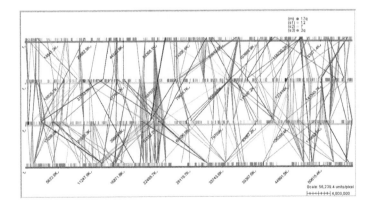

FIGURE 9.7   Comparison of multiple chromosomes in Bluejay. Human chromosomes 17q, 12, 7, and 2q are compared in Bluejay, which shows that many genes are duplicated on several chromosomes, as represented by the many lines that link genes belonging to the same gene family.

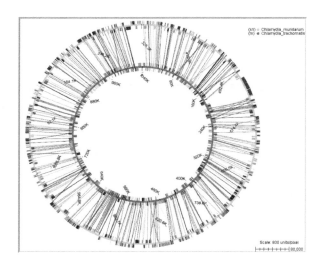

FIGURE 9.8   Comparison of two bacterial genomes in Bluejay in circular representation. Comparing *Chlamydia trachomatis* and *Chlamydia muridarum* reveals that they have many common genes according to the Gene Ontology (GO) classification, as shown by the linking lines. Only those links with an angular distance less than 2% of the maximum possible distance (360 degrees) are shown.

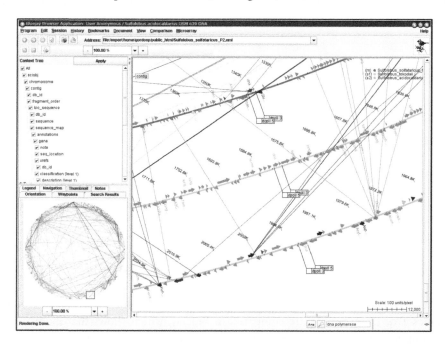

FIGURE 9.9   Genome alignment by waypoints in Bluejay. A waypoint named "dpoII" is set at the appropriate location in each of three *Sulfolobus* spp. genomes. On selecting "Align at Waypoint" operation, the genomes are aligned at the dpoII gene by appropriately rotating the two outer genomes. This facilitates easy visual comparison of the gene's structure in all three species.

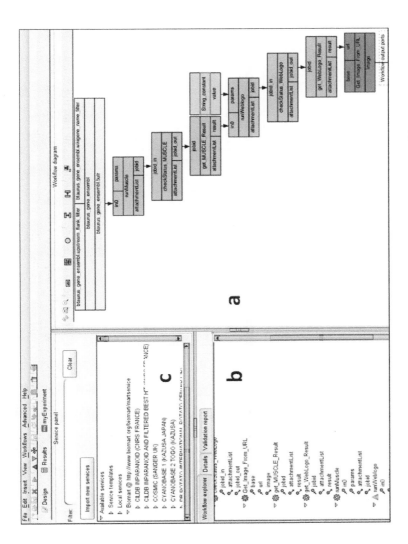

FIGURE 10.5 Taverna interface in Design mode. Main components are: (a) interactive graphical workflow diagram; (b) workflow explorer, hierarchical listing of workflow inputs, workflow outputs, processors with their ports, and data links between ports; and (c) list of available processors for inclusion in the workflow.

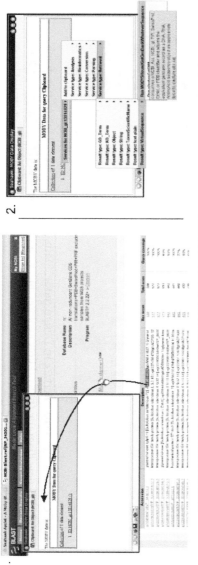

FIGURE 10.8 Seahawk service browsing: From BLAST Web page to sequence and species name tabs in Seahawk. The steps are (1) drag a DB identifier from a Web browser onto the Seahawk clipboard to import the ID; (2) click the ID link and select a service to retrieve the record; (3) Shift+Click the AminoAcidSequence result and select a cross-reference utility (result opens in new tab due to Shift+Click); (4) click the taxon ID to get the species name; and (5) view the result.

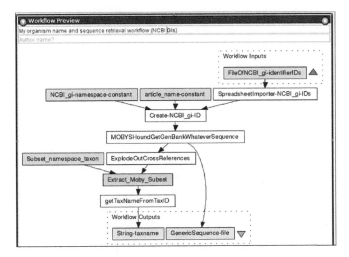

FIGURE 10.9   Seahawk's workflow preview for the Figure 10.8 analysis demonstration.

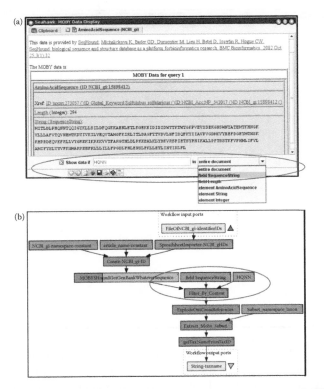

FIGURE 10.10   Seahawk's conditional service execution via a search/filter widget. (a) Filter conditions in Seahawk GUI and (b) corresponding filtering processors in a running Taverna workflow.

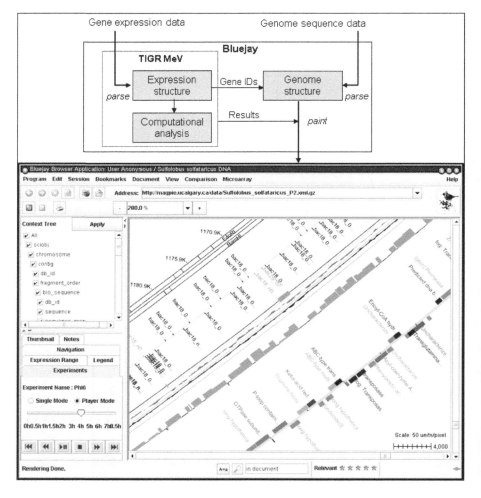

FIGURE 11.4   Integration of gene expression analysis with genomic data in Bluejay. Gene expression data parsing and analysis are done by TIGR MeV. The expression and analysis data are combined with the genome data to produce a unified visual representation, where the genes and their expression/analysis values are displayed side by side.

CHAPTER 7

Functional Annotation

7.1 INTRODUCTION

For those genes that are translatable into proteins, and their surrounding neighborhoods, a number of analyses can be performed to link the proposed messenger RNA (mRNA) constructs (see Chapter 2) to their predicted biological functionality. As these annotations are mostly one-time assignments (such as the predicted gene function of a certain locus within the genome), they are usually performed by using "static" annotation tools (see Chapter 8). We define static annotation tools as those that create a mapping of features to genomic regions, with tables, spreadsheets, or HTML-based Web pages as the final output. Typically, users can interfere little or not at all with the final output of these systems and have to take the mappings created by the static annotation tools at face value. This may be problematic, as the public data repositories contain a large number of erroneous entries, which might lead to incorrect functional assignments.

Primarily, functional predictions are based on similarity searches of the nucleic acid data and the resulting amino acid translations of the open reading frames, against public data, which are stored in repositories such as the EMBL database, GenBank, or more specialized data collections as outlined in this chapter.

7.2 BIOPHYSICAL AND BIOCHEMICAL FEATURE PREDICTION

7.2.1 Physical Chemistry Features

The distribution of hydrophobic (i.e., water-hating) amino acids in the predicted protein can indicate whether a protein segment is likely to be naturally located in solution next to or within a (fatty) membrane. For example, the EMBOSS (Rice et al., 2000) Pepwheel program calculates a so-called helical wheel to check the periodicity of hydrophobic residues. With the proper periodicity, one side of a protein alpha helix is hydrophobic and the other hydrophilic (i.e., water-loving). This would indicate that the protein might be a transmembrane (TM) transporter, or some other protein interacting with both fatty membranes and aqueous molecules. More generally, the prediction of transmembrane protein segments is done using amino acid distribution models based on known transmembrane genes. Early success in transmembrane prediction was achieved by using hidden Markov models (HMMs) (Sonnhammer et al., 1998). Subsequently, methods were developed to refine this technique. Current best practice is to combine multiple TM predictions. For example, the results of six different transmembrane predictors are combined in MetaTM (Klammer et al., 2009) using a statistical method known as support vector machines (Burges, 1998). These methods may also indicate whether the N- or C-terminus of the protein is expected to be outside of the membrane. A protein may contain multiple transmembrane domains in which case the inner-outer prediction is made for each nonmembrane segment.

7.2.2 Sequence Motif Prediction

Small amino acid sequence patterns, known as motifs, may not have biochemical functions themselves in the cell but may indicate that a protein is the target of various biochemical or transport processes in the cell.

7.2.2.1 Protein Modification

Amino acids in a protein may undergo chemical modification to change or regulate their biological function. More than 750 types of posttranslation modifications (PTMs) have thus far been documented (Montecchi-Palazzi et al., 2008). Some of the most common PTMs are the addition of an acetyl group (CH_3CO, acetylation), formic acid (HCOOH, formylation), a phosphate group (PO_4^{3-}, phosphorylation), a polysaccharide (glycosylation), or an ubiquitin protein (ubiquitination) (Parker et al., 2010). Among other things, phosphorylation is important for the activation of proteins involved in cell signaling (Krebs and Beavo, 1979). Glycosylation affects protein solubility, stability, protein half-life, and immunogenicity (Dennis et al., 1999). Ubiquitination has several roles, including a key role in protein recycling in eukaryotes (Komander, 2009). Simple amino acid motifs for common sites, such as those found in the PROSITE database (Hulo et al., 2007), can generate many false positives and false negatives. Another approach is to build tools specifically for a particular class of PTM, for example, kinase phosphorylation (Xue et al., 2006). PHOSPHIDA (Gnad et al., 2011) provides a comprehensive database and prediction tools for the three modifications already mentioned and therefore is more generally useful. The dbPTM database (Lee et al., 2006) uses experimentally anno-tated PTM sites to build HMMs for more than 20 common PTM classes. These HMMs can be downloaded for local use. AutoMotifServer (Basu and Plewczynski, 2010) provides an integrated platform with machine learning to predict the most common modifications with confidence, and it can be used either online or installed locally.

Prediction of PTMs can also be important to help explain the observa-tion of unusual peptide masses when mass spectrometry (MS) data are being mapped back to an annotated genome. It is computationally expen-sive to incorporate every possible modification into an MS search, there-fore prior annotation of PTM predictions with a high confidence level can speed up such searches considerably.

7.2.2.2 Protein Localization

Motifs in the amino acid sequence can cause a protein to be directed toward specific parts of the cell. These motifs are known as localization signals. Signal peptides that direct proteins to be secreted from the cell can be identified using SignalP (Petersen et al., 2011) for all organisms and Phobius (Käll et al., 2007) for eukaryotes only. These tools can be used to match the N-terminus of the protein to a sequence model, which was derived using known secreted proteins. The two tools can also be used to predict membrane-anchored proteins. TargetP (Emanuelsson et al., 2007) augments SignalP to allow predictions of the presence of N-terminal motifs for chloroplast transit peptides and mitochondrial-targeting peptides.

Programs that combine multiple predictions and inspect both the N- and C-protein termini achieve more comprehensive localization analysis for prokaryotes. PSORTb (Yu et al., 2010) is a prime example, combining infor-mation on signal peptides, targeting motifs, and transmembrane topologies. LocateP (Zhou et al., 2008) is another combination predictor with accuracy in the vicinity of 90%. In prokaryotes LocateP's options for primary local-izations are C-terminally anchored, intracellular, lipid anchored, LPxTG cell-wall anchored, multitransmembrane, N-terminally anchored, secreted via minor pathways (bacteriocin), and secretory (released).

LOCtree (Nair and Rost, 2005) is based on separate decision trees specific to plants, nonplant eukaryotes, and prokaryotes. In prokaryotes it can be used to classify secreted, periplasmic, and cytoplasmic proteins. In eukaryotes proteins are assigned to one of six localizations: extracellular, nuclear, cytoplasmic, chloroplastic, mitochondrial, or other organelles. BaCelLo (Pierleoni et al., 2006) implements different predictors for the three main eukaryotic kingdoms: animals, plants, and fungi. The composition of sequence profiles derived from lineage-specific multiple sequence alignments can strongly affect the scoring. Proteins are designated as one of secreted, nuclear, cytoplasmic, mitochondrial, or chloroplast origin.

The annotation of localization signals can be used to gain valuable insights into cell adaption. For example, a localization signal may be contained within an exon that is missing in some splice variants (see Section 2.6). This would indicate differential trafficking of a protein based on external factors. Localization can also provide support to other forms of functional annotation. For example, genes annotated as carbohydrate degradation enzymes in fungal genomes should generally contain signal peptides, since these enzymes are known to be secreted (Braaksma et al., 2010).

7.3 PROTEIN DOMAINS

The functionality of a protein is largely determined by its three-dimensional structure, which itself can be broken down into functional subunits, also known as protein domains. These protein domains conserve three-dimensional structure and charge distributions. Protein domains facilitate diverse biochemical functions, such as metal ion binding or luminescence. Proteins can contain several domains, for example, a carbohydrate binding domain to juxtapose a complex sugar (the ligand) next to the protein and a carbohydrate degradation domain to subsequently cleave the ligand. Tens of thousands of conserved domains are known to exist, based on the size of the InterPro catalog (Hunter et al., 2011). The restrictions on structure and charge to maintain functionality are reflected in the rate of amino-acid substitution in each position of the secondary (linear amino acid) sequence. Position-specific scoring models (PSSMs) are used to capture these domain constraints and are referred to as protein domain models. The most commonly used model is an HMM (Krogh et al., 1994). As illustrated in Figure 7.1, the model represents the totality of positive examples given in the aligned input. This allows more sensitive matching of a query to the model than to any particular positive example.

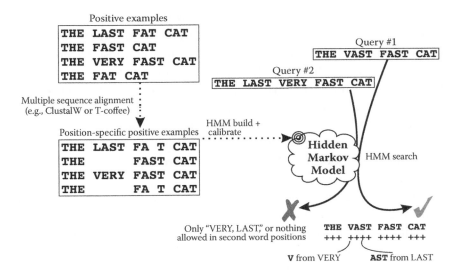

FIGURE 7.1 An example of how a hidden Markov model (HMM) is built and how the matching is not pairwise (the first query instance matches even though VAST does not appear in the training set) as well as being position specific (the second instance does not match, because the LAST and VERY are mutually exclusive in the training set).

The net effect of this sensitive, functional subunit matching is twofold. First, recently created protein sequences, which are being annotated, may share little secondary sequence similarity to publicly available sequences, but if overall domain constraints are preserved, these sequences can often still be functionally identified. Second, a general function can be inferred from a small conserved domain match, even if the rest of the protein sequence has no similarity to known genes. For example, a protein may contain a DNA-binding domain at the C-terminus. *DNA binding protein* could therefore be used as the general functional annotation of the gene if no evidence points to a more specific function, for example, a transcription factor or restriction endonuclease.

Protein sequences can be compared to an HMM using the HMMer software (Finn et al., 2011) or a hardware accelerator such as the TimeLogic DeCypher system (Active Motif, Carlsbad, California). Version 3 of HMMer forgoes the exhaustive Viterbi algorithm (Forney, 1973) used to optimally match the HMM and instead opts for a few assumptions to achieve 100-fold speedup in search time. A caveat of HMMer3 (Eddy, 2011) is that the heuristics that allow this speedup can also generate matches to partial domains with significant e-values. As these domains are incomplete, they are most likely nonfunctional. Older HMMer searches and Viterbi alignments, such as those found in the toolkit included in the DeCypher system, will not report partial model traversals. In some situations, the partial matching may be useful, such as the identification of truncated gene prediction. For example, if the C-terminus of a predicted protein contained a strong partial match to the N-terminal domain of glycerol kinase, it is likely that the predicted start and stop codon for the gene are incorrect or the genomic DNA sequence has been misassembled.

Alternative methods for modeling domain constraints have also been implemented. RPS-BLAST (Altschul et al., 1997) can be used to search specially formatted position-specific scoring matrices, which are created from PSI-BLAST-derived models of protein domains. PROSITE (Hulo et al., 2007) provides both Gribskov profiles (similar to HMMs but less probabilistic) and regular expressions to model smaller protein domains. The PRINTS database (Attwood et al., 2003) employs a search for multiple, closely spaced areas of high conservation, ignoring areas a larger HMM might model loosely.

The construction of domain models, regardless of the representation used, typically involves the automated detection of potential positive examples, followed by the manual curation of these examples. The model statistics are then calibrated as necessary to reduce false positives, while minimizing false negatives. Annotation of the function of the protein domain is also generally a manual process. Many different models could be derived for the same protein domain, depending on the source for positive examples, the underlying PSSM mechanism used, and the subsequent manual curation steps. Several domain model databases exist, and large ones such as Pfam (Finn et al., 2010) and PANTHER (Thomas et al., 2003) have somewhat overlapping and somewhat distinct model sets.

In order to unify these data collections, the InterPro Consortium (www.ebi.ac.uk/interpro) was created to correlate models from all of the main providers, assigning related models to named families and using a common identifier system. It is important to note that InterPro does not contain any domain models, rather it points to the models of the individual providers. As illustrated in Figure 7.2, InterProScan (Zdobnov and Apweiler, 2001) simplifies analysis by providing a unified search result from the various HMM, RPS-BLAST, and PRINTS searches.

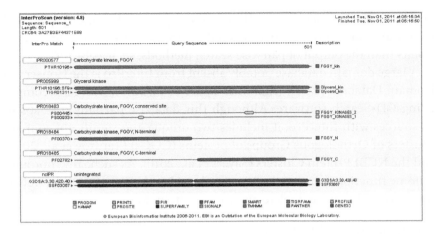

FIGURE 7.2 Unified, visual protein domain output from InterProScan, showing the different ways in which the same domain can be modeled.

In the example shown in Figure 7.2, two models are grouped together in each of three categories. The identifiers on the left of each match line indicate the source database of the model. For instance, in the InterPro Match IPR005999 (Glycerol kinase), PTHR10196:SF9 is from the PANTHER database (Thomas et al., 2003), while the similar TIGR01311 is from TIGRFAM (Haft et al., 2003). The interpretation of these InterProScan results requires some subtle observations. First, clearly different domain models provide different coverage of the 501 amino acid query protein. In some cases, the N- and C-termini are modeled separately. In other cases (PROSITE motifs PS00445 and PS00933 in IPR018483), only a very small and highly conserved region is modeled. Second, it is not necessarily obvious that all of the InterPro categories (noted by IPR numbers) are interrelated. Is this a possibly multifunction gene? Good background knowledge or traversal of the IPR ID hierarchy is necessary to discover that all of these IPR terms fall into a tight branch of the protein domain hierarchy. Third, which is the most specific functional description that can be reasonably assigned to the query gene? Examining the text output from InterProScan is necessary, as this lists the random expectations (e-values) for the domain matches. Glycerol kinase is the most specific term in the hierarchy, and the e-value for this functional assignment is 0, indicating that this is a very strong match: *glycerol kinase* is therefore a reasonable annotation for the query protein. In cases where e-values are closer to 1, the interpretation is more difficult, especially because the different model sources do not necessarily share the same statistical model. Despite these caveats, protein domain matches still tend to provide more sensitive, less redundant and often better-annotated results than the output of pairwise search methods, such as BLAST.

A large domain database notably absent from InterPro is the Conserved Domain Database (CDD) (Marchler-Bauer et al., 2011), with its accompanying CD-Search software. Although this database partially overlaps in its sources with InterPro, it includes two unique and significant sources: Clusters of Orthologous Groups of proteins (COGs) (Tatusov et al., 2003) and the NCBI Protein Clusters (Sayers et al., 2010). As such, any thorough genome functional annotation system should make use of CDD as well as InterPro to obtain a complete picture of protein domains.

7.4 SIMILARITY SEARCHES

By far the most common method for functional annotation is based on similarity searches. The premise of this method is that a high degree of pairwise nucleic acid and/or amino acid conservation between an already characterized and annotated gene sequence and a query gene sequence also implies conservation of the function, which was determined for the first gene sequence. This simple premise is complicated to implement in practice, because similarity is in many cases more or less a continuum, rather than a yes–no trait. While each piece of software will use its own e-value guidelines on homology detection for functional annotation (see Chapter 8), the definition of homologous sequences is based on the evolutionary origin of the related genes and not on simple sequence similarity. Key types of homologs are described next according to these origins.

7.4.1 Paralogs

Any organism may contain several similar genes that originate from a single instance of the gene, which has undergone one or more duplication events during evolution. Over time, these genes tend to specialize in their function, and might be expressed in specific tissues or might show ligand specificity (Guillén et al., 2010). If one instance of such a gene sequence has been already characterized, it is usually reasonable to infer a similar function for the other members of the gene family based on very strong amino acid level pairwise similarity (e.g., e-value < 10^{-35}). In terms of nomenclature, typically paralogs constitute a gene family with similar gene symbols, such as *hox1, hox2,* and *hox3.* HOX in particular is a large, ancient paralog family, predating modern chordate evolution, and it is therefore challenging to trace the duplication sequence precisely (Hughes et al., 2001). Naturally, the nearer a duplication event is to the leaves of the evolutionary tree, the stronger the degree of nucleotide similarity between the paralogous sequences should be.

Ohnologs are a special case of paralogs, as they are thought to be derived at the same time from a double whole-genome duplication event in the common chordate ancestor (Leveugle et al., 2003). This common origin means all of these gene families have roughly the same evolutionary distance. Yeast geneticists have also adopted the term ohnolog, as several yeast species have undergone whole genome duplication, but evolution rates of ohnologs are not equivalent in these lower species (Bu et al., 2011).

7.4.2 Orthologs

Strictly speaking, two genes from different species are called orthologs if they are each other's best hit in a pairwise comparison of genes between the two species (best reciprocal hit), possibly supplemented by maximum likelihood distance data in the case of confounding paralogs (Wall et al., 2003). The genes are more closely related to each other than anything else, therefore they will likely share a common evolutionary origin. In general, orthologs will have the same major biological function, meaning that an existing annotation can be reused at a coarse level. Gene symbols for orthologs are often shared among species, so that, for example, the ING1 gene in the human genome is the equivalent of the ING1 gene in the *Drosophila* genome.

It should be noted though that there are several exceptions to the aforementioned principles. Orthologs can have different names and numbers depending on when they were originally annotated, for example, ING1's ortholog in yeast is called YNG1, and ING2 is also YNG1's ortholog (Gordon et al., 2008). It is also possible that orthologs may play a functionally critical role in one organism but not in another. For example, 20% of the human orthologs that have been determined to play a critical role in the mouse do not have the same critical function in humans (Liao and Zhang, 2008). Orthologous transcription factors in bacteria may target different recognition sites (Price et al., 2007). Even the one-to-one relationship between orthologs may not always be universal. In Figure 7.3, the tree-based gene comparison of eukaryotic VNN1 genes from Ensembl (Flicek et al., 2010) shows that a recent gene duplication event occurred in the African elephant (*Loxodonta africana*) genome. Due to the very high degree of sequence similarity, both of these genes could be considered orthologs of the human VNN1 gene.

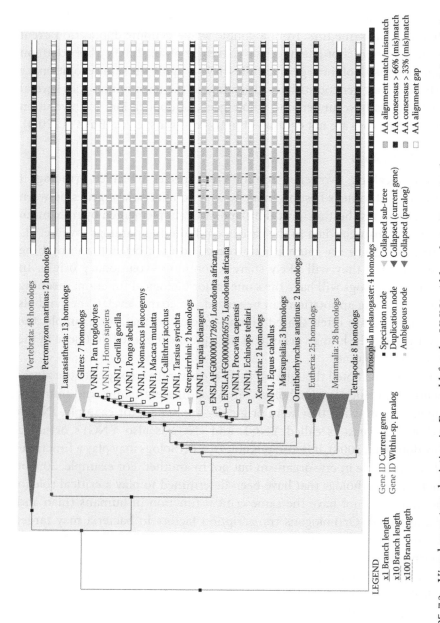

FIGURE 7.3 Visual gene tree depiction in Ensembl for the VNN1. Although most species have one copy of the orthologous gene, *Loxodonta africana* has two, showing that orthologs are not necessarily one-to-one. **(See color insert.)**

Because Ensembl's tree-based method is computationally expensive, other techniques and data sets for the comprehensive determination of orthologous sequences have been developed. The InParanoid ortholog database (Ostlund et al., 2010) includes over 100 species and bases orthology on a two-pass BLAST approach while using special scoring matrices. OrthoMCL (Chen et al., 2006) includes nearly as many species, but uses a Markov clustering algorithm. High-throughput local computation of orthologs can be achieved using OrthoMCL or Ortholuge (Fulton et al., 2006). The latter uses a pairwise-comparison, supplemented by statistical analysis relative to remotely related ("outgroup") genes.

7.4.3 Xenologs

Genes are not necessarily only related via a common ancestor or a shared gene duplication event. In many instances, genes can also be acquired by an organism from another species via a horizontal or lateral gene transfer (HGT or LGT) event, which is most frequently observed in prokaryotes and protists (Andersson, 2005). Although the gene will evolve independently in the new host organism, a strong sequence similarity will exist between the genes in the donor and the recipient organisms. Famously, archaeal genes are the best match for about 11% of the bacterial *Thermotoga maritima* genome (Nelson et al., 1999), suggesting that a significant portion of the genome has been laterally acquired. HGT is known to occur between almost all branches of life, with varying frequency. Annotation of xenologs can be important to determine such instances, because newly acquired genes will conflict with most assumptions that an annotator has in terms of homology-based annotation, such as orthology in related species.

A database of precomputed xenologs for many species is available (Garcia-Vallve et al., 2003). Methods for the detection of xenologs are primarily based on nucleotide sequence identity and relics of the old genome organization, such as codon usage. These are supplemented by various statistical approaches, such as Bayesian networks (Needham et al., 2007) and support vector machines (Burges, 1998). During the identification of genes that might have been introduced into an organism by lateral gene transfer, every method used must make a trade-off between false positives and false negatives. A promising recent development is a "multithreshold" approach (Azad and Lawrence, 2011), which first filters clearly atypical alien genes and clearly typical native genes, using conservative thresholds for parameters such as the base composition. Ambiguous gene sequences are subsequently classified with less strict thresholds, after an examination of the nature of their flanking genes (native or alien).

7.4.4 Analogs

Analogs are genes that do not share a clear common evolutionary origin but that perform similar biological functions, sometimes in a host–pathogen relationship (Sawitzke and Stahl, 1992). These gene sequences may contain similar protein domains based on the principle of convergent functional evolution (which of course leads to similar three-dimensional structures for similar gene function and thus requires similar amino acid distributions). In some cases where a gene has not been characterized, an analog may show up as the best match with a functional description.

7.5 PAIRWISE ALIGNMENT METHODS

Determination of homologs requires the use of pairwise sequence alignment methods. The available methods vary in their utility depending on many factors, such as uniqueness of the species under study and sequence quality. There are two main measures of the successfulness of an alignment search:

1. *Sensitivity (i.e., recall)*—What proportion of the real hits are reported? (More sensitive means more real hits.)

2. *Selectivity (i.e., precision)*—What proportion of the reported hits are real? (More selective means less false positives.)

The classic trade-off in alignment searches is speed versus thoroughness. The speed and thoroughness of common alignment methods are examined here.

7.5.1 Canonical Methods

The Needleman–Wunsch algorithm (Needleman and Wunsch, 1970) was the first theoretically optimal method to align two DNA sequences. Matches receive positive scores, while mismatches and gaps receive negative scores. The method produces a global alignment, that is, from the first to last base of both the query and target. A global alignment tends to introduce numerous gaps between truly homologous sequences, lowering the alignment score significantly, for example, when the gene length has changed (i.e., by inserting terminal gaps). Smith and Waterman (1981) produced an algorithm to find the best local alignment by adjusting the scoring of gaps in the optimal solution derivation. This method produces alignment scores (and related statistics such as Z scores) suitable for comparing more distantly related homologous sequences. The Smith–Waterman method was further refined by others (Altschul and Erickson, 1986) to mimic real protein evolution more closely; inserting a gap in a sequence is costly, but extending the same gap is less so. This is known as affine gap alignment. Computing the optimal alignment solution, using a technique known as dynamic programming (Eddy, 2004), requires large-scale computation. Recently, specialized software implementations (Rognes, 2011) of the Smith–Waterman algorithm take advantage of the extended instruction sets of modern CPUs. This allows the execution of Smith–Waterman searches in roughly the same amount of time as heuristic methods that have traditionally been used for alignment practicality.

7.5.2 Heuristic Methods

Pairwise alignment methods such as BLAST are called "heuristic," because they achieve a speedup in database search through assumptions about what the final alignments will look like. The bias is generally toward moderately to well-conserved sequences without frameshifts and possibly with gaps that do not greatly affect the overall alignment length. When searching for strong homologs, these assumptions are very reasonable, but because many shortcuts are taken there are actually many more parameters to BLAST than to the canonical Smith–Waterman method. To make the least number of assumptions about what the matches will look like, the Smith–Waterman method searches through a lot more data, being an exhaustive method. Tweaking the parameters of the BLAST tools can increase their sensitivity. This naturally lengthens the runtime of the searches.

Key parameters to BLAST for finding weak homologs are the word size, step size, gap score, and X drop-off score. The word size determines the minimum exact match length that must be present to "anchor" the initial alignment. For DNA, the default minimum is 11 base pairs, while amino acid alignments require 3 minimum residues. Particular to amino acid alignments is a "neighborhood score," which defines how many conservative amino acid substitutions should be tried in the anchors as well. The higher the neighborhood score, the more remote matches will be found. The step size parameter determines what fraction of possible anchors is examined. For example, setting the step size to 1 ensures every 11-mer in the query is tried (if the default word size of 11 is used), and setting the step size to 2 checks every second anchor. The sum of the word size and the step size determines the lower boundary; alignments with longest exact match smaller than this may not be found.

Once initial alignment anchors are found, ungapped extension of the alignment occurs. The X drop-off score determines how much this extension tolerates mismatches: the higher the value, the longer extensions will be. The final phase of the alignment allows extension and recalculation of the ungapped alignment using a method much more akin to the canonical Smith-Waterman algorithm. Setting the gap open and gap extension penalties lower than the default, particularly for DNA alignments, will allow longer sequences with more distant relationships between the sequences to be properly aligned.

Conversely, any or all of these parameters may be tightened in order to reduce BLAST runtime. Other popular heuristic methods such as BLAT (Kent, 2002) and SSAHA (Ning et al., 2001) can be used to speed up BLAST-like alignments for single genome targets by concentrating even further on primarily identifying highly homologous sequences.

7.5.3 Scoring Matrices

Regardless of whether canonical or heuristic methods are used to do pairwise alignment, every method requires a scoring matrix to determine the match and mismatch values to be applied to the sequences. For amino acid alignments in particular, the scoring matrix determines which amino acid substitutions are considered conservative (and hence score highly) and which are considered deleterious to function (and hence scored negatively). Two matrix series commonly used in bioinformatics are BLOSUM (Henikoff and Henikoff, 1992) and PAM (Schwartz and Dayhoff, 1978). The matrices in these series are numbered; for example, BLOSUM62 is the default scoring matrix for BLAST amino acid searches. The BLOSUM series of matrices was constructed by looking at substitution rates of amino acids in un-gapped blocks, which are contained within multiple sequence alignments of many protein families. The number of the matrix (e.g., 62 in BLOSUM62) indicates the percentage identity that was required in the block to consider it. As such, BLOSUM matrices with higher numbers model substitutions in closely related proteins better. For example, BLOSUM90 is based on ungapped blocks from proteins with at least 90% identity. In the PAM matrix series, the number indicates the number of random point mutations introduced into every 100 residues of test sequences before net effects on amino acid substitution were calculated. In this case, smaller numbers model substitutions in closely related proteins better (i.e., PAM100 is more conservative than PAM200).

The matrix used for pairwise alignment affects the raw alignment score, which in turn determines the e-value of the alignment. In cases where homology is being detected between closely related species (e.g., human and chimpanzee), more conservative substitution matrices will yield a more accurate alignment e-value. Specialty matrices can also be applicable in particular cases. In the case of distant homology, the OPTIMA (Kann et al., 2000) matrix provides improved pairwise alignment sensitivity. PHAT (Ng et al., 2000) and SLIM (Müller et al., 2001) provide better modeling of transmembrane protein substitution rates. This is because the BLOSUM and PAM series are derived from globular protein data sets, which by their nature contain far fewer hydrophobic residues than membrane proteins (which reside in a lipid environment).

7.6 CONCLUSION

Functional annotation is somewhat based on biochemical features of the predicted amino acid sequences of genes, and largely based on similarity to other proteins. Similarity can be based on overall alignment or domain-specific features. Pairwise alignments and HMMs are normally used to make these determinations, respectively. The choice of method and scoring mechanism for pairwise alignment can affect the sensitivity of homology detection and hence the quality of functional annotation.

REFERENCES

Altschul, S.F., Erickson, B.W. 1986. Optimal sequence alignment using affine gap costs. *Bull. Math. Biol.* 48:603–616.

Altschul, S., Madden, T., Schaffer, A., et al. 1997. Gapped BLAST and PSI-BLAST: A new generation of protein database search programs. *Nucleic Acids Res.* 25:3389–3402.

Andersson, J.O. 2005. Lateral gene transfer in eukaryotes. *Cellul. Mol. Life Sci.* 62:1182–1197.

Attwood, T.K., Bradley, P., Flower, D.R., et al. 2003. PRINTS and its automatic supplement, prePRINTS. *Nucleic Acids Res.* 31(1):400–402.

Azad, R.K., Lawrence, J.G. 2011. Towards more robust methods of alien gene detection. *Nucleic Acids Res.* 39(9):e56.

Basu, S., Plewczynski, D. 2010. AMS 3.0: prediction of post-translational modifications. *BMC Bioinformatics* 11:210.

Braaksma, M., Martens-Uzunova, E.S., Punt, P.J., Schaap, P.J. 2010. An inventory of the *Aspergillus niger* secretome by combining *in silico* predictions with shotgun proteomics data. *BMC Genomics* 11:584.

Bu, L., Bergthorsson, U., Katju, V. 2011. Local synteny and codon usage contribute to asymmetric sequence divergence of Saccharomyces cerevisiae gene duplicates. *BMC Evol. Biol.* 11:279.

Burges, C.J.C. 1998. A tutorial on support vector machines for pattern recognition. *Data Min. Knowl. Discov.* 2:121–167.

Chen, F., Mackey, A.J., Stoeckert, C.J. Jr., Roos, D.S. 2006. OrthoMCL-DB: Querying a comprehensive multi-species collection of ortholog groups. *Nucleic Acids Res.* 34:D363–D368.

Dennis, J.W., Granovsky, M., Warren, C.E. 1999. Protein glycosylation in development and disease. *Bioessays* 21:412–421.

Eddy, S.R. 2004. What is dynamic programming? *Nat. Biotechnol.* 22:909–910.

Eddy, S.R. 2011. Accelerated profile HMM searches. *PLoS Comput. Biol.* 7:e1002195.

Emanuelsson, O., Brunak, S., von Heijne, G., Nielsen, H. 2007. Locating proteins in the cell using TargetP, SignalP, and related tools. *Nat. Protoc.* 2:953–971.

Finn, R.D., Clements, J., Eddy, S.R. 2011. HMMER Web server: Interactive sequence similarity searching. *Nucleic Acids Res.* 39:W29–W37.

Finn, R.D., Mistry, J., Tate, J., et al. 2010. The Pfam protein families database. *Nucleic Acids Res.* 38:D211–D222.

Flicek, P., Aken, B.L., Ballester, B., et al. 2010. Ensembl's 10th year. *Nucleic Acids Res.* 38(Database issue):D557–D562.

Forney, G.D. Jr. 1973. The Viterbi algorithm. *Proc. IEEE* 61:268–278.

Fulton, D.L., Li, Y.Y., Laird, M.R., Horsman, B.G., Roche, F.M., Brinkman, F.S. 2006. Improving the specificity of high-throughput ortholog prediction. *BMC Bioinformatics* 7:270.

Garcia-Vallve, S., Guzman, E., Montero, M.A., Romeu, A. 2003. HGT-DB: A database of putative horizontally transferred genes in prokaryotic complete genomes. *Nucleic Acids Res.* 31:187–189.

Gnad, F., Gunawardena, J., Mann, M. 2011. PHOSIDA 2011: The post-translational modification database. *Nucleic Acids Res.* 39:D253–D260.

Gordon, P.M., Soliman, M.A., Bose, P., Trinh, Q., Sensen, C.W., Riabowol, K. 2008. Interspecies data mining to predict novel ING-protein interactions in human. *BMC Genomics* 9:426.

Guillén, D., Sánchez, S., Rodríguez-Sanoja, R. 2010. Carbohydrate-binding domains: multiplicity of biological roles. *Appl. Microbiol. Biotechnol.* 85:1241–1249.

Haft, D.H., Selengut, J.D., White, O. 2003. The TIGRFAMs database of protein families. *Nucleic Acids Res.* 31(1):371–373.

Henikoff, S., Henikoff, J.G. 1992. Amino acid substitution matrices from protein blocks. *Proc. Natl. Acad. Sci. USA* 89:10915–10919.

Hughes, A.L., da Silva, J., Friedman, R. 2001. Ancient genome duplications did not structure the human Hox-bearing chromosomes. *Genome Res.* 11:771–780.

Hulo, N., Bairoch, A., Bulliard, V., et al. 2007. The 20 years of PROSITE. *Nucleic Acids Res.* 36:D245–D249.

Hunter, S., Jones, P., Mitchell, A., et al. 2011. InterPro in 2011: New developments in the family and domain prediction database. *Nucleic Acids Res.* 40:D306–D312.

Käll, L., Krogh, A., Sonnhammer, E.L. 2007. Advantages of combined transmembrane topology and signal peptide prediction—The Phobius Web server. *Nucleic Acids Res.* 35:W429–W432.

Kann, M., Qian, B., Goldstein, R.A. 2000. Optimization of a new score function for the detection of remote homologs. *Proteins* 41:498–503.

Kent, W.J. 2002. BLAT—The BLAST-like alignment tool. *Genome Res.* 12:656–664.

Klammer, M., Messina, D.N., Schmitt, T., Sonnhammer, E.L. 2009. MetaTM—A consensus method for transmembrane protein topology prediction. *BMC Bioinformatics* 10:314.

Komander, D. 2009. The emerging complexity of protein ubiquitination. *Biochem. Soc. Trans.* 37:937–953.

Krebs, E.G., Beavo, J.A. 1979. Phosphorylation-dephosphorylation of enzymes. *Annu. Rev. Biochem.* 48:923–959.

Krogh, A., Brown, M., Mian, I.S., Sjölander, K., Haussler, D. 1994. Hidden Markov models in computational biology. Applications to protein modeling. *J. Mol. Biol.* 235:1501–1531.

Lee, T.Y., Huang, H.D., Hung, J.H., Huang, H.Y., Yang, Y.S., Wang, T.H. 2006. dbPTM: An information repository of protein post-translational modification. *Nucleic Acids Res.* 34:D622–D627.

Leveugle, M., Prat, K., Perrier, N., Birnbaum, D., Coulier, F. 2003. ParaDB: A tool for paralogy mapping in vertebrate genomes. *Nucleic Acids Res.* 31:63–67.

Liao, B.Y., Zhang, J. 2008. Null mutations in human and mouse orthologs frequently result in different phenotypes. *Proc. Natl. Acad. Sci. USA* 105(19):6987–6992.

Marchler-Bauer, A., Lu, S., Anderson, J.B., et al. 2011. CDD: A Conserved Domain Database for the functional annotation of proteins. *Nucleic Acids Res.* 39:D225–D229.

Montecchi-Palazzi, L., Beavis, R., Binz, P.A., et al. 2008. The PSI-MOD community standard for representation of protein modification data. *Nat. Biotechnol.* 26:864–866.

Müller, T., Rahmann, S., Rehmsmeier, M. 2001. Non-symmetric score matrices and the detection of homologous transmembrane proteins. *Bioinformatics* 17(Suppl. 1):S182–S189.

Nair, R., Rost, B. 2005. Mimicking cellular sorting improves prediction of subcellular localization. *J. Mol. Biol.* 348:85–100.

Needham, C.J., Bradford, J.R., Bulpitt, A.J., Westhead, D.R. 2007. A primer on learning in Bayesian networks for computational biology. *PLoS Comput. Biol.* 3:e129.

Needleman, S.B., Wunsch, C.D. 1970. A general method applicable to the search for similarities in the amino acid sequence of two proteins. *J. Mol. Biol.* 48:443–453.

Nelson, K.E., Clayton, R.A., Gill, S.R., et al. 1999. Evidence for lateral gene transfer between Archaea and bacteria from genome sequence of *Thermotoga maritima*. *Nature* 399:323–329.

Ng, P.C., Henikoff, J.G., Henikoff, S. 2000. PHAT: A transmembrane-specific substitution matrix. *Bioinformatics* 16:760–766.

Ning, Z., Cox, A.J., Mullikin, J.C. 2001. SSAHA: A fast search method for large DNA databases. *Genome Res.* 11:1725–1729.

Ostlund, G., Schmitt, T., Forslund, K. et al. 2010. InParanoid 7: New algorithms and tools for eukaryotic orthology analysis. *Nucleic Acids Res.* 38:D196–D203.

Parker, C.E., Mocanu, V., Mocanu, M. 2010. Mass spectrometry for post-translational modifications. In *Neuroproteomics*, ed. O. Alzate, 93–114. Boca Raton, FL: CRC Press.

Petersen, T.N., Brunak, S., von Heijne, G., Nielsen, H. 2011. SignalP 4.0: Discriminating signal peptides from transmembrane regions. *Nat. Methods* 8(10):785–786.

Pierleoni, A., Martelli, P.L., Fariselli, P., et al. 2006. BaCelLo: A balanced subcellular localization predictor. *Bioinformatics* 22:e408–e416.

Price, M.N., Dehal, P.S., Arkin, A.P. 2007. Orthologous transcription factors in bacteria have different functions and regulate different genes. *PLoS Comput. Biol.* 3(9):e175.

Rice, P., Longden, I., Bleasby, A. 2000. EMBOSS: The European Molecular Biology Open Software Suite. *Trends Genet.* 16(6):276–277.

Rognes, T. 2011. Faster Smith-Waterman database searches with inter-sequence SIMD parallelisation. *BMC Bioinformatics* 12:221.

Sawitzke, J.A., Stahl, F.W. 1992. Phage lambda has an analog of *Escherichia coli* recO, recR and recF genes. *Genetics* 130:7–16.

Sayers, E.W., Barrett, T., Benson, D.A., et al. 2010. Database resources of the National Center for Biotechnology Information. *Nucleic Acids Res.* 38:D5–D16.

Schwartz, R.M., Dayhoff, M.O. 1978. Matrices for detecting distant relationships. In *Atlas of Protein Sequence and Structure,* ed. M.O. Dayhoff, 353–358. Washington, DC: National Biomedical Research Foundation.

Smith, T., Waterman, M.S. 1981. Identification of common molecular subsequences. *J. Mol. Biol.* 147:195–197.

Sonnhammer, E.L.L., von Heijne, G., Krogh, A. 1998. A hidden Markov model for predicting transmembrane helices in protein sequences. *Proc. Int. Conf. Intell. Syst. Mol. Biol.* 6:175–182.

Tatusov, R.L., Fedorova, N.D., Jackson, J.D., et al. 2003. The COG database: An updated version includes eukaryotes. *BMC Bioinformatics* 4:41.

Thomas, P.D., Campbell, M.J., Kejariwal, A., et al. 2003. PANTHER: A library of protein families and subfamilies indexed by function. *Genome Res.* 13(9):2129–2141.

Wall, D.P., Fraser, H.B., Hirsh, A.E. 2003. Detecting putative orthologs. *Bioinformatics* 19:1710–1711.

Xue, Y., Li, A., Wang, L., Feng, H., Yao, X. 2006. PPSP: Prediction of PK-specific phosphorylation site with Bayesian decision theory. *BMC Bioinformatics* 7:163.

Yu, N.Y., Wagner, J.R., Laird, M.R., et al. 2010. PSORTb 3.0: Improved protein subcellular localization prediction with refined localization subcategories and predictive capabilities for all prokaryotes. *Bioinformatics* 26(13):1608–1615.

Zdobnov, E.M., Apweiler, R. 2001. InterProScan—An integration platform for the signature-recognition methods in InterPro. *Bioinformatics* 17(9):847–848.
Zhou, M., Boekhorst, J., Francke, C., Siezen, R.J. 2008. LocateP: Genome-scale sub-cellular-location predictor for bacterial proteins. *BMC Bioinformatics* 9:173.

CHAPTER 8

Automated Annotation Systems

8.1 INTRODUCTION

After database searches, any given gene in a genome may have some or all of the functional evidence listed in Chapter 7. Before complete genomes were available, the search results were mostly inspected and interpreted manually. With the advent of complete microbial (and later eukaryotic) genomes, tools had to be created that processed, filtered, and displayed these types of information in an automated way. Many such tools now exist. In general, static automated genome annotation systems perform the following steps:

- Organization of the genomic information (e.g., number of contigs, and grouping of contigs that are derived from the same chromosome)

- Identification of biophysical and biochemical properties

- Database searches of the potentially coding genomic regions and the complete genomic sequence

- Organization of the search results in tabular format and potentially in XML format

- Creation of a Web interface that provides access to the results

In this chapter, we are focusing on three of these tools, which can handle both prokaryotic and eukaryotic genome annotations, as well as one tool that was specifically designed for the annotation of microbial genomes generated using next-generation DNA sequencing methods. The examples described here are typical in their design and function for the class of static genome annotation systems. Various approaches to job management, data filtering, and user interfaces in genome annotation are highlighted. These design principles and concepts are also applicable to many other genome annotation tools that are not specifically mentioned here.

8.2 MAGPIE

MAGPIE is an automated system for the structural and functional annotation of genomes and transcriptomes. Its origins (Gaasterland and Sensen, 1996) are in the sequencing of the *Sulfolobus solfataricus* P2 archaeal genome (She et al., 2001), but it has constantly evolved and has been applied to numerous archaeal, bacterial, and eukaryotic data sets. It runs on any UNIX-like system and is available from http://sourceforge.net/projects/magpie09/. An overview of the major components of the annotation system is depicted in Figure 8.1.

8.2.1 Analysis Management

In any large-scale annotation system, a management system for tasks is a must. MAGPIE used a customized version of the simple Autoson queuing system (http://hdl.handle.net/1885/40815). This system allows jobs to be run in parallel, on multiple machines, and in a given order, for example, BLAST (Altschul et al., 1997) homolog searches before InterPro (Hunter et al., 2011) domain searches. Priorities can be assigned to different analysis steps to ensure rapid access to the most relevant results. Specific hosts can be defined for specific jobs based on required resources (e.g., memory intensive jobs go to large-memory computers). Configuration of the available hosts and their priorities is managed through a simple text, while a small number of UNIX shell commands allow the user to pause, resume, and delete jobs as well as recreate failed ones. Pending analyses can be viewed through the MAGPIE Web interface. Privacy settings for the Web interface are controlled via standard Apache Web server (http://*httpd. apache.org)* access files. Data is stored in indexed flat files, meaning no database expertise is required.

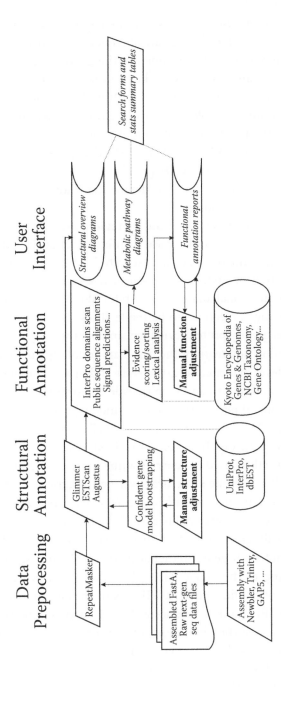

FIGURE 8.1 Overview of the major components comprising the MAGPIE functional genomic annotation system. Manual intervention steps (in bold) are optional.

8.2.2 Structural Annotation

Before starting annotation, a genome can be run through a repeat masking program (see Chapter 3) to clean up repetitive and possibly misassembled genome regions. After importing the genome sequence file, the first step in genome annotation is predicting the protein coding sequence (CDS) (see Chapter 2). For prokaryotes with high guanine (G) and cytosine (C) (G+C) content, simple open-reading frame (ORF) finding can be sufficient due to the low false-positive rate (stop codons are rich in adenine (A) and thymine (T) [A+T]). For most other prokaryotes, imported Glimmer (Delcher et al., 2007) CDS predictions are sufficient. Intergenic regions are subsequently scanned to determine if any small ORFs with strong homologs were overlooked.

For eukaryotes, potential protein coding regions are initially determined by searching a public protein database (configurable) against the genome with a double-affine (intron-spanning) search. Each region's segments are spliced together to form a putative messenger RNA (mRNA), which is subjected to protein functional analysis. A search is done for mRNAs that have no apparent splicing corrections to be made. These have a length that is a multiple of three, quality protein evidence, an in-frame stop codon, and an in-frame stop codon upstream of the first start codon upstream of the best quality protein-level evidence (i.e., the start codon is unambiguous). The search for starts and stops can creep outside the original mRNA prediction into the genome if necessary. These "confident genes" are used to train ESTScan (Iseli et al., 1999) (CDS only and UTRs) and then AUGUSTUS (whole genomic structure of the gene). The trained AUGUSTUS (Stanke et al., 2006) data files are then used to generate a final gene prediction. Figure 8.2 shows the MAGPIE genome structural overview diagram, where depicted for each gene is the protein coding sequence extent (box arrow) and the mRNA extent (lines with terminal bars, with blanks in the introns).

The overview diagram quickly indicates several important pieces of information about each gene predicted in the 1.02 Mbp sequence, with a legend at the top of the image. First, the color of the box arrows indicates the degree of confidence MAGPIE has in the automated functional annotation. MAGPIE ranks evidence at level 1 (strong domains or likely orthologs found), 2 (other homolog found), or 3 (marginal evidence). Genes assigned to these levels are colored blue, cyan, or gray respectively. Second, whether start and stop codon locations are determined yet is indicated by the outline of the box arrow. Predicted genes with in-frame start

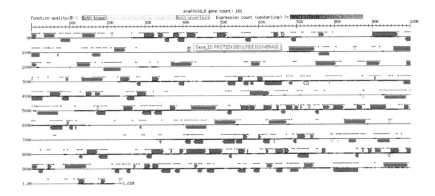

FIGURE 8.2 MAGPIE genomic sequence structural overview. Each arrow box represents a gene on either strand. The broken lines above these show the exon structure of each transcript, along with demarcations of the gene starts and stops (vertical cap lines). Mapped RNA-Seq read density is shown in red (in varying brightness) on the centerline. Element and outline coloration is explained in the legend at the top of the images. **(See color insert.)**

and stop codons have a black outline. A start or stop codon may be missing because of misassembly of the genome or a poor statistical model for parts of the gene, in which case the outline color is not black.

A third type of evidence that can be displayed in the overview is transcriptional evidence. Transcript data such as next-generation sequencing of mRNA can be mapped back to the genome and stored in the SAM/BAM format (Li et al., 2009). When such a mapping file is available, MAGPIE displays the depth of mapped transcript data by the color of centerline around which positive and negative strand genes fall. A brighter red color indicates higher read depth. Taken all together, the overview quickly lets a human annotator see which genes have strong evidence (protein or transcript), which need start or stop codon adjustment, and which have unexpectedly long or short mRNAs. This information can guide any effort that needs to be put into the manual adjustment of the gene structure. As shown in Figure 8.2, setting the mouse over a gene reveals a tooltip with the calculated functional description for the gene (here "protein disulfide isomerase"). Clicking the gene takes the user to the functional annotation display and confirmation form, described in Section 8.2.3.

Figure 8.3 shows the interface for manually adjusting mRNA structure in MAGPIE. In this example the original mRNA model, spanning eight exons, is broken into two separate genes by clicking "Create New Gene

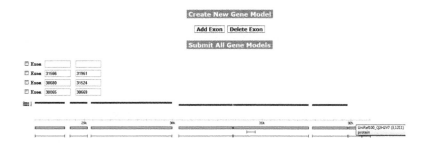

FIGURE 8.3 MAGPIE interface for editing or adding gene transcript models. Coordinates can be entered manually in the text boxes (top) or populated automatically by selecting homology matches (bottom).

Model," adding three exons, then clicking "Create New Gene Model" again, and adding the remaining exons. The coordinates of each exon are populated by clicking the thin red evidence bars or manually entering them. Setting the mouse over an evidence bar shows a tooltip summarizing the evidence (lower right corner of the figure).

So-called chimeric genes, where two protein products are predicted as one, can be caused by relying on poorly constructed protein models in the public databases. As high-throughput genomics yields more automated gene information in the public databases, it is important to look at the totality of public protein evidence rather than just the top pairwise hit. This is where structural and functional annotation is intertwined. While mRNAs must be manually split, in "confident" CDS extraction, MAGPIE examines all the top protein hits to determine if, minus one or two, the hits fall into sets with nonoverlapping locations. If this is the case, separate coding sequences are predicted, with the suffixes .1, .2, and so on.

8.2.3 Functional Annotation

Predicted coding sequences undergo functional analysis as described in Chapter 7. As pictured in Figure 8.4, the results are combined to generate a functional evidence summary report that can be used to confirm or adjust the automated functional assignment. These data are typical of the sorts of information shown in other functional annotation systems. The elements in the summary will be explained top to bottom.

The first line ("MAGPIE Suggestion") indicates the synthesized functional description MAGPIE generates from the totality of evidence collected. Many other annotation systems simply use the best BLAST or protein domain hit description, but this can often be uninformative

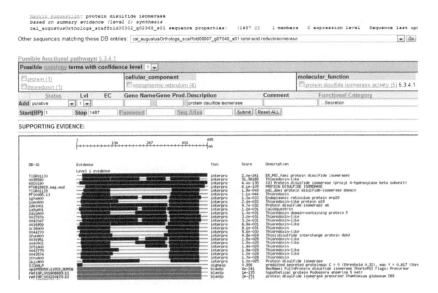

FIGURE 8.4 MAGPIE functional annotation form. From top to bottom: a textual description synthesized from all of the evidence, a list of possible paralogs/homologs, biochemical pathway information, Gene Ontology terms, an edit facility for the functional annotations, start/stop editing, and analysis evidence. Clicking the analysis evidence shows the original tool output, whereas clicking the ID leads to the public database entries. **(See color insert.)**

or unrepresentative. MAGPIE assigns weights to the words found in all the evidence and then only reports those with strong weights in the order they are usually found in the individual hits (bottom of Figure 8.4). The weight of a word is based on its frequency, the evidence rank, and in the case of protein, matches the taxonomic distance between the homolog organism and the one being annotated. Additional postprocessing indicates if there is ambiguity in the assignment such as paralog number (e.g., alcohol dehydrogenase 1/2). Tissues, species names, database identifiers, and uninformative words such as "putative," "hypothetical," or "unknown" are stripped. Custom lists of these "stop" words can be provided, along with "kill" words that will cause an entire hit to be ignored. Kill words can be useful, for example, when reannotating a public genome where the existing public annotation should be ignored.

The synthesized functional description is followed by a dropdown list of other genes in the genome that match the same kinds of public data. This can help the user navigate between related genes such as paralogs and

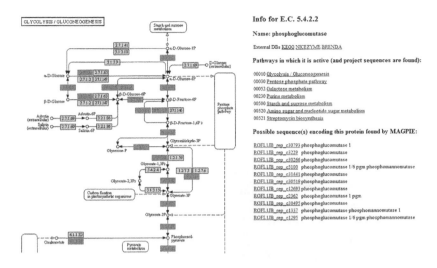

FIGURE 8.5   A MAGPIE pathway diagram (left) based on the KEGG database. Clicking a red EC number (e.g., 5.4.2.2 at the top of the pathway) links to an enzyme summary specific to this genome.

general homologs. Below the dropdown is a link to the biochemical pathways in which the putative gene may be involved ("Possible functional pathways: 5.3.4.1"). Enzyme Commission (EC) numbers (e.g., 5.3.4.1) (Bairoch, 2000) are derived from available EC mappings from UniProt, NCBI, and InterPro. MAGPIE generates EC summary pages, an example of which is shown on the right part of Figure 8.5, based on available information from the KEGG database (Kanehisa et al., 2012). The generic biochemical pathways are then regenerated with genes available in this genome highlighted. The user can navigate between various genes and pathways using this map. If most genes' ECs are not highlighted, the pathway may not be active in the organism under study. If only one or two are missing, it may indicate functional annotations that need manual intervention.

Manual annotation can be entered in the various input elements below the EC link. The first set of predictions that can be confirmed or denied are Gene Ontology (GO; www.geneontology.org) terms. These terms are derived from lexical analysis of the hits and additional information, such as available mapping files for UniProt ID. GO terms are a common way to annotate genes because it allows the grouping and comparison of genes across one or more species at different levels of granularity, without needing exact matching of functional description text. Following the GO terms are text boxes where the automated functional description and confidence level can be manually overridden. The annotator would base such an edit on the evidence presented in the graph underneath. Each database hit can be clicked to see the original sequence alignment. Hits are grouped into level 1, 2, and 3 as described earlier. Within a level, the order of the hits is by tool trustworthiness (configurable), then score, then length, then alphabetical order. Where the match is to a protein sequence, the phylogeny of the sequence is indicated by the "DB-ID" background color; this can help the annotator quickly identify xenologs. The color key is as follows:

- Viruses—Brown

- Bacteria—Green

- Archaea—Yellow

- Eukarya (other than those below)—Pink

- Fungi—Orange

- Plants—Purple

- Metazoa—Red

From the protein domain, signal, and protein hits in the graph, it can be determined that this is likely a secreted protein disulfide isomerase from fungi. Manual functional annotation is not a requirement, rather it provides an opportunity for the user to correct borderline automated calls of function or start and stop codon. This is an acknowledgment that no automated system will perfectly capture existing biological knowledge. It also allows new information for particular genes being studied in the lab to be included and shared with colleagues.

8.2.4 User Interface

Several MAGPIE Web interface pages have been shown so far, which can be browsed from the genome summary image by an annotator. Quite frequently, a user will want to not browse or edit but rather search the genome annotation. The main search interface for MAGPIE is shown in Figure 8.6.

The first search option is a full text search of analysis results, with a filter for the quality of the evidence. This provides a more sensitive way to find sequences than the typical top descriptions indexed by other annotation Web sites. Second, a user may search for genes assigned a given GO term (or its subterms). The interface dynamically pulls up a list of

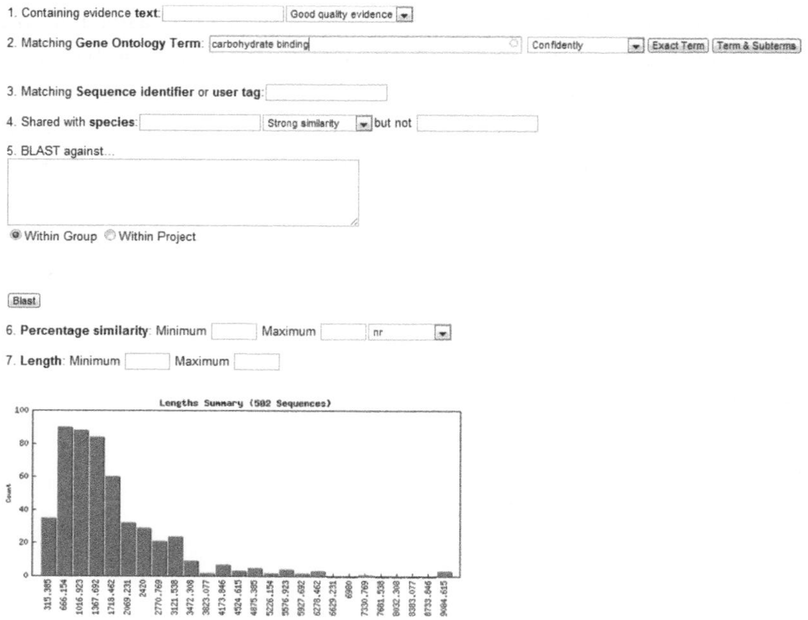

FIGURE 8.6 MAGPIE annotation search form. Ways to search, from top to bottom, are free text, Gene Ontology, sequence ID, taxonomy, custom sequence similarity, sequence similarity in collected evidence, and sequence length.

GO terms matching free text entered by the user. This allows the use of terms even if the user is not familiar with the exact terminology in GO. This can also capture genes that may not be described using the plain text labels the user expects or are too general to be captured by a few keywords. For example, capturing all carbohydrate metabolic genes with a few keywords is impossible, but the "carbohydrate metabolism" GO term is likely assigned to many genes in the genome.

The third way to search is using the automatically generated MAGPIE ID for a gene or a user-generated nametag. These custom nametags can be more memorable and can be set in the functional annotation form ("Seq Alias" in Figure 8.4).

The fourth search type, unique to MAGPIE, is by taxonomic distribution of the gene. This search allows the user to find all lineage-specific genes, universal genes, or recent xenologs. The taxonomy names can be from any level of the NCBI Taxonomy database (www.ncbi.nlm.nih.gov/Taxonomy), including common names. For example, in the search for drug target candidates for a human pathogenic bacterium, one might want to exclude genes with strong similarity to "Homo sapiens," "human" or "9606." Any of these terms would map to the appropriate taxonomy filter. This filter is made possible by maintaining an index of the NCBI taxonomy and its mapping to public protein IDs.

The fifth type of search is sequence-based. The user may have the DNA or protein sequence of a gene of interest from another organism. A sequence search is a guaranteed way to find related genes in the new genome regardless of the functional annotation results. Conversely the sixth search type, specifying a minimum or maximum match similarity against public databases, can be used to quickly screen for novel (or well-known) genes. Finally, the seventh search allows the user to focus on very long or very short genes as desired.

In summary, MAGPIE provides all of the key features required for structural and functional annotation of a genome, including synthesis of data from many analysis tools, and with extensive facilities to interrogate the results. The implementation of these features in other systems will be overviewed to show how different approaches can be taken for the same tasks.

8.3 GENERIC MODEL ORGANISM DATABASE

The Generic Model Organism Database (GMOD; http://gmod.org) is a well-known community-driven effort to build reusable components for the storage, analysis, and display of genomic information. Within the large set of software tools GMOD makes available is an automated annotation system called MAKER (Cantarel et al., 2008). MAKER can be used for annotation in prokaryotic and smaller eukaryotic genome projects, creating genome databases that can be used by other GMOD components.

8.3.1 Analysis Management

Command-line based like MAGPIE, MAKER uses an internal job management system, the text logs of which the user can review. Data is also stored in flat files except when there are more than 1000 sequences in the project. In this case a custom data store contains the analysis results. Job parallelization on a single machine is achieved by running the analysis launcher several times in a row or by configuring the system's Message Passing Interface (MPI).

8.3.2 Structural Annotation

MAKER first identifies repeats in the genome and masks them. It then aligns ESTs and proteins to the masked genome. By default, SNAP (Korf,

2004) produces *ab initio* gene predictions, but other predictors mentioned in Chapter 2 can be used as well. Each potential exon is scored according to several criteria, some of which can be configured:

- Length of the 5′ and 3′ untranslated regions (UTRs)

- Confirmation of splice sites by expressed sequence tag (EST) alignment

- Overlap of EST and protein alignments with the *ab initio* prediction

- Total number of exons and total protein length

High-scoring exons are merged into a gene annotation with quality values. To improve prediction accuracy, MAKER can also be bootstrapped. The outputs of preliminary runs can be used to automatically retrain the gene prediction algorithm. Such an iterative process requires some diligence and manual postanalysis to determine when to stop iterating the models and which configuration parameters to modify to correct systematic prediction biases.

8.3.3 Functional Annotation

The MAKER-generated model transcripts are first translated into protein FastA files. At a basic level, MAKER maps putative functions identified from BLASTP against UniProt/SwissProt (Magrane et al., 2011). Beyond this "best hit" approach, mechanisms to do functional annotation using GMOD components are not standardized. GMOD components are used by many model organism communities, each of which has their own preferred mechanisms for generating functional annotation. The variety of methods they use reflect the different data source challenges faced by plant, fungal, insect, mammal, and other communities, as well as the amount of human and computational resources they have available to devote to the annotation process. The commonality of these systems is that they store their results in the GMOD database system, called Chado (Zhou et al., 2006). From a Chado database, several GMOD tools can then further process the data, for example, for visualization.

8.3.4 User Interface

When run on the command line, MAKER's inputs are essentially a FastA-formatted sequence file and transcript data if available. A Web-based version of MAKER is available that requests several of the

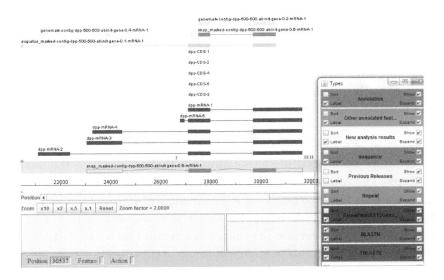

FIGURE 8.7   Apollo genome browser view of MAKER genomic structural annotation. A consensus gene structure is displayed at the bottom (light blue background color), with contributing evidence stacked on top (different colors for
different evidence sources according to the legend at right). Apollo allows graphical editing of model borders, with write-out to GFF file format. **(See color insert.)**

aforementioned inputs to be uploaded to the server. MAKER outputs
are in GFF3 or FastA format, and are intended to be loaded directly into
GMOD genome browser components. The GBrowse viewer (Stein et al.,
2002) can overlay the annotation information along with other data
sources for viewing. For editing the annotations, the Apollo genome
browser can be used. An example is shown in Figure 8.7, where Apollo
displays the final MAKER prediction along with all of the evidence collected to generate it.

8.4 AGeS

As sequencing bacterial genomes has become routine, the need for fast,
automated annotations has spurred the development of many tools for
their analysis. AGeS (Kumar et al., 2011) is a typical example of a prokary-
otic-focused genome annotation tool.

8.4.1 Analysis Management

AGeS uses the OpenMPI library (www.open-mpi.org) to achieve parallelization of compute-intensive tasks, and manages the scheduling of tasks

via the PBS (www.pbsworks.com) queueing system. The state of analysis tasks and their results are stored in an embedded relational database management system (RDBMS) called Apache Derby (http://db.apache.org), which requires little manual administration. This database works in conjunction with the Web server to provide a graphical interface for the job monitoring and results display of the genome annotation process.

8.4.2 Structural Annotation

AGeS builds on an existing annotation pipeline called DIYA (Stewart et al., 2009). DIYA was customized to predict gene locations using Glimmer (Delcher et al., 2007), ribosomal RNAs using RNAmmer (Lagesen et al., 2007), and transfer RNAs using tRNAscan-SE (Lowe and Eddy, 1997). The simplicity of gene detection in prokaryotes means that a single method of structural prediction can be sufficient for most purposes.

8.4.3 Functional Annotation

The functional annotation part of AGeS is called PIPA. An automated ontology mapping generation algorithm maps various classification schemes into the GO. PIPA predicts protein functions by mapping common GO terms to the disparate results of multiple programs and databases, such as InterPro and the NCBI's Conserved Domains Database discussed in Chapter 7. A protein profile generation algorithm creates customized profile databases to predict specific protein functions, in order to improve annotation specificity. The algorithm was employed to construct the built-in enzyme profile database CatFam (Yu et al., 2009), which predicts catalytic functions described by EC numbers.

The final consensus annotation is based on an automated reconciliation of the integrated programs and databases. Calculating the agreement between various prokaryotic annotation pipelines can be done at multiple levels. A common method is to compare the EC or GO term mappings. Using this methodology, AGeS annotations agree highly with those generated by popular Web-based systems such as IMG (Markowitz et al., 2010). The major differences between prokaryotic annotation systems tend to be in the functional description text and the degree of linking to external data sources.

8.4.4 User Interface

Whereas some prokaryotic pipelines have an integrated user interface for results display, the output of AGeS (and many other pipelines) is a

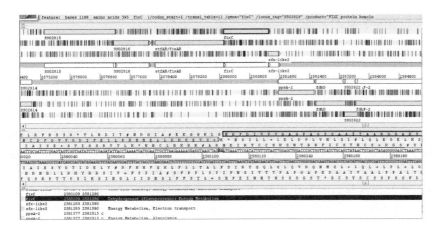

FIGURE 8.8 Artemis browser view of an archaeal GenBank file. (Top) Six-translation frame genomic structural overview (with stop codon lines). (Middle) Genomic DNA sequence and six-frame translation. (Bottom) GenBank feature entries (editable).

GenBank-formatted text file. These files can be loaded into annotation editors such as Artemis (Carver et al., 2008), which can overlay additional data sources onto the annotation to correct it. Figure 8.8 shows a screenshot of Artemis. GenBank-file-based editors allow the user to modify, add, or delete features at the level described by the GenBank format, as at the bottom of the figure. Because prokaryotic genes tend to be considerably smaller than eukaryotic ones, the DNA sequence and six-frame translation can be reasonably shown on the screen in addition to the high-level structural view.

8.5 ENSEMBL

The Ensembl Web site of annotated public genomes (Flicek et al., 2010) is a well-used resource for genome biology, provided by the European Bioinformatics Institute. GMOD and Ensembl have some crossover; they both export their annotation information to the jointly developed BioMart software (Zhang et al., 2011). BioMart provides fast, intuitive queries over these genome-scale data sets. The Ensembl annotation tools can be downloaded for local installation, but GMOD remains the more common toolset used locally by species-specific communities with emerging genomes. The key strengths of Ensembl lies in its scalability, variety of integrated analyses, and user interface.

8.5.1 Analysis Management

Ensembl's codebase contains a set of Perl modules (Potter et al., 2004) that coordinate analysis tasks across machines, known as Hive. Hive interfaces to the commercial cluster computing tools Load Sharing Facility and Sun Grid Engine to execute analyses. Input to and results from command-line analyses are shuttled between files and a mySQL database (www.mysql.com). The database has a particular data schema shared by all Ensembl instances. Given that the public Ensembl Web site manages a huge number of genomes and query traffic, load management of Web-based access is controlled at the server clusters level using commercial information technology (IT) software (e.g., the Amazon Elastic Cloud Compute environment). For individual genomes though, the database setup and interface are engineered to serve a larger eukaryotic genome on a single computer.

8.5.2 Structural Annotation

Ensembl has two main methods for providing structural annotation: for high- and low-quality genome assemblies. In the case of emerging (low coverage) genomes, genes may span multiple assembly contigs and may also contain a significant number of sequencing errors. In order to predict full gene structures, a BLAST-like search tool for very long alignments (Schwartz et al., 2003) is employed against the human genome. These alignments are fed to a reference sequence scaffolder (Kent et al., 2003) and custom scripts to ensure nonredundancy of exon assignments. Putative frameshifts in the exon sequences are corrected using the human reference. As such, low-coverage structural annotation is done completely by homology.

Structural annotation of high-quality assemblies employs many more steps. Before annotation begins, the genome is uploaded to a mySQL database according to the Ensembl database schema. Repeat masking and (species-dependent) *ab initio* gene predictors are then run. The *ab initio* predictions are not used as part of the structure determination steps that follow but provide an additional source of information for manual curation.

Of primary importance to structural annotation is protein homology. The first stage of the structural annotation process is referred to as the targeted stage. When available, known protein sequences from the organism under study are aligned to the genome using the intron-spanning GeneWise alignment algorithm (Birney et al., 2004) with strict identity parameters. The next stage, called the similarity stage, applies the same alignment technique with somewhat less stringent parameters to the regions of the genome not mapped by the targeted stage. The relative contribution of the two stages is dependent on the amount of existing public data available for the species and its close relatives.

The second part of structural prediction is cDNA alignment. Species-specific cDNA (assembled) and EST (single-pass) transcript sequences are aligned to the genome. If a cDNA alignment overlaps gene regions predicted in the protein-based stages, the nonoverlapping region from the cDNA alignment is annotated as predicted 5′ or 3′ UTR. EST alignments have considerably less weight as supporting evidence in the annotation process due to their low quality.

Finally, a nonredundant set of transcript models for a gene is produced by merging identical transcripts derived from different mapped proteins. All of the final transcript models are presented to the user in the final interface, some or all of which may represent actual biological transcripts.

8.5.3 Functional Annotation

A number of postprocessing steps filter and annotate the predicted genes. Ensembl accumulates evidence from tools mentioned in Chapter 7, such as SignalP and related tools (Emanuelsson et al., 2007), and InterPro (Hunter et al., 2011). These data are displayed in the protein product summary page, as shown in Figure 8.9.

Noncoding RNA annotation is done using Rfam (Gardner et al., 2011). A check is done for a starting methionine residue and a trailing polyadenylation signal (see Chapter 2). In eukarya, pseudogenes can be tricky to distinguish. Ensembl's pseudogene prediction is based on matching one of the following criteria:

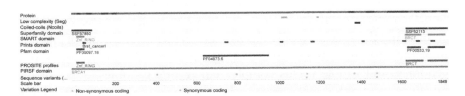

FIGURE 8.9 Ensembl functional evidence summary. (Top) Exon structure, with alternating light and dark purple to show boundaries. (Middle) Two biochemical feature and six domain homology lines of evidence. (Bottom) Known sequence variants (when available). (**See color insert.**)

- The transcript has a single exon and matches a multiexon transcript elsewhere.

- RepeatMasker identifies the transcript as completely repetitive.

- The transcript has multiple frameshifts and no introns.

- The transcript has frameshifts and all introns are at least 80% repetitive.

Ensembl does not synthesize a gene description but rather cross-links information available from existing protein annotation data sources such as UniProt and the NCBI Gene databases (www.ncbi.nlm.nih.gov/gene). Manual structural and functional curation of Ensembl genes can be done using Apollo (Lewis et al., 2002). Unique among genome annotation tools, Ensembl incorporates several of the genome-associated data types discussed in Chapter 4 to predict genomic regulatory features as well. Empirical data from functional genomics experiments include:

- Chromatin immunoprecipitation—Tiling arrays (ChIP-chip), sequencing, and DNase sequencing

- Epigenomic modifications—Histone, transcription factors, RNAPollII, DNA methylation, data from the ENCODE project

- Regulatory motifs

- MicroRNA (miRNA) targets

8.6 SUMMARY

Functional annotation is essential to understanding emerging genomes. Functional annotation is of course predicated on good structural annotation. Genome annotation pipelines exist for handling these two stages and automating the collection of bioinformatics evidence. A variety of pipeline implementations exist in order to target prokaryotic versus eukaryotic genomes, and simple versus complex eukaryotic genomes. Some of the pipelines are stand-alone, whereas others are meant to be used in conjunction with other components for editing or viewing the automated annotation results. Chapter 9 explores several examples of these annotation data display and editing environments.

REFERENCES

Altschul, S., Madden, T., Schaffer, A., et al. 1997. Gapped BLAST and PSI-BLAST: A new generation of protein database search programs. *Nucleic Acids Res.* 25:3389–3402.

Bairoch, A. 2000. The ENZYME database in 2000. *Nucleic Acids Res.* 28(1):304–305.

Birney, E., Clamp, M., Durbin, R. 2004. GeneWise and Genomewise. *Genome Res.* 14:988–995.

Cantarel, B.L., Korf, I., Robb, S.M., et al. 2008. MAKER: An easy-to-use annotation pipeline designed for emerging model organism genomes. *Genome Res.* 18(1):188–196.

Carver, T., Berriman, M., Tivey, A., et al. 2008. Artemis and ACT: Viewing, annotating and comparing sequences stored in a relational database. *Bioinformatics* 24:2672–2676.

Delcher, A.L., Bratke, K.A., Powers, E.C., Salzberg, S.L. 2007. Identifying bacterial genes and endosymbiont DNA with Glimmer. *Bioinformatics* 23(6):673–679.

Emanuelsson, O., Brunak, S., von Heijne, G., Nielsen, H. 2007. Locating proteins in the cell using TargetP, SignalP, and related tools. *Nat. Protoc.* 2:953–971.

Flicek, P., Aken, B.L., Ballester, B., et al., 2010. Ensembl's 10th year. *Nucleic Acids Res.* 38(Database issue):D557–D562.

Gaasterland, T., Sensen, C.W. 1996. MAGPIE: Automated genome interpretation. *Trends Genet.* 12(2):76–78.

Gardner, P.P., Daub, J., Tate, J., et al. 2011. Rfam: Wikipedia, clans and the "decimal" release. *Nucleic Acids Res.* 39(Database issue):D141–D145.

Hunter, S., Jones, P., Mitchell, A., et al. 2011. InterPro in 2011: New developments in the family and domain prediction database. *Nucleic Acids Res.* 40:D306–D312.

Iseli, C., Jongeneel, C.V., Bucher, P. 1999. ESTScan: A program for detecting, evaluating, and reconstructing potential coding regions in EST sequences. *Proc. Int. Conf. Intell. Syst. Mol. Biol.* 138–148.

Kanehisa, M., Goto, S., Sato, Y., Furumichi, M., Tanabe, M. 2012. KEGG for integration and interpretation of large-scale molecular data sets. *Nucleic Acids Res.* 40(Database issue):D109–D114.

Kent, W.J., Baertsch, R., Hinrichs, A., Miller, W., Haussler, D. 2003. Evolution's cauldron: duplication, deletion, and rearrangement in the mouse and human genomes. *Proc. Natl. Acad. Sci. USA* 100:11484–11489.

Korf, I. 2004. Gene finding in novel genomes. *BMC Bioinformatics* 5:59.

Kumar, K., Desai, V., Cheng, L., et al. 2011. AGeS: A software system for microbial genome sequence annotation. *PLoS One* 6(3):e17469.

Lagesen, K., Hallin, P., Rødland, E.A., Staerfeldt, H.H., Rognes, T., Ussery, D.W. 2007. RNAmmer: Consistent and rapid annotation of ribosomal RNA genes. *Nucleic Acids Res.* 35:3100–3108.

Lewis, S.E., Searle, S.M.J., Harris, N., et al. 2002. Apollo: A sequence annotation editor. *Genome Biol.* 3:RESEARCH0082.

Li, H., Handsaker, B., Wysoker, A., et al. 2009. The Sequence Alignment/Map format and SAMtools. *Bioinformatics* 25:2078–2079.

Lowe, T.M., Eddy, S.R. 1997. tRNAscan-SE: A program for improved detection of transfer RNA genes in genomic sequence. *Nucleic Acids Res.* 25:955–964.

Magrane, M., Consortium, U. 2011. UniProt Knowledgebase: A hub of integrated protein data. *Database* 2011:bar009.

Markowitz, V.M., Chen, I.M., Palaniappan, K., et al. 2010. The integrated microbial genomes system: An expanding comparative analysis resource. *Nucleic Acids Res.* 38:D382–D390.

Potter, S.C., Clarke, L., Curwen, V., et al. 2004. The Ensembl analysis pipeline. *Genome Res.* 14(5):934–941.

Schwartz, S., Kent, W.J., Smit, A., et al. 2003. Human-mouse alignments with BLASTZ. *Genome Res.* 13(1):103–107.

She, Q., Singh, R.K., Confalonieri, F., et al. 2001. The complete genome of the crenarchaeon *Sulfolobus solfataricus* P2. *Proc. Natl. Acad. Sci. USA* 98:7835–7840.

Stanke, M., Keller, O., Gunduz, I., Hayes, A., Waack, S., Morgenstern, B. 2006. AUGUSTUS: *Ab initio* prediction of alternative transcripts. *Nucleic Acids Res.* 34(Web Server issue):W435–W439.

Stein, L.D., Mungall, C., Shu, S., et al. 2002. The generic genome browser: A building block for a model organism system database. *Genome Res.* 12:1599–1610.

Stewart, A.C., Osborne, B., Read, T.D. 2009. DIYA: A bacterial annotation pipeline for any genomics lab. *Bioinformatics* 25(7):962–963.

Yu, C., Zavaljevski, N., Desai, V., Reifman, J. 2009. Genome-wide enzyme annotation with precision control: Catalytic families (CatFam) databases. *Proteins* 74(2):449–460.

Zhang, J., Haider, S., Baran, J., et al. 2011. BioMart: A data federation framework for large collaborative projects. *Database* 2011:bar038.

Zhou, P., Emmert, D., Zhang, P. 2006. Using Chado to store genome annotation data. *Curr. Protoc. Bioinformatics* Chapter 9:Unit 9.6.

Dynamic Annotation Systems

End-User-Driven Annotation and Visualization

9.1 INTRODUCTION

Recent advances in DNA sequencing and analysis technologies have resulted in a proliferation of sequence and annotation data in overwhelming varieties of data types and information granularities. One of the central challenges in genome annotation is the interpretation and visualization of large amounts of genomic data generated from static annotation pipelines. To study complex biological entities such as genomes with scientific annotations and associated data, a user-oriented computational tool that helps the user to visualize the genome annotations and integrate the data with their own results is required. Ideally, such a visually enriched environment allows researchers to explore a genomic sequence at multiple viewing scales and in customized ways, to enable them to derive meaningful interpretations of their annotation data. An image-based software system for viewing and integrating genomic data needs to provide the user with the best possible imagery that elucidates the data in a unified visual context. Essentially, this represents the second generation of genome analysis and annotation systems, which is much more flexible especially on the graphical level as compared to the earlier static genome annotation systems.

There are a number of software tools that are loosely classified as genome browsers that can be used to display genome data and biological annotations using a graphical user interface. Examples include the University of California, Santa Cruz (UCSC) Genome Browser (Kent et al., 2002), Ensembl Genome Browser (Flicek et al., 2011), NCBI Map Viewer (www.ncbi.nlm.nih.gov/mapview), GBrowse (Stein et al., 2002), Bluejay (Soh et al., 2012; Soh et al., 2008), NCBI Genome Workbench (www.ncbi.nlm.nih.gov/projects/gbench), Integrated Genome Browser (Nicol et al., 2009), and Apollo (Lewis et al., 2002). Some of these are Web-based, whereas others are locally installable tools, as described in Sections 9.2 and 9.3, respectively.

The annotated data are usually organized into display units in the form of linear or circular (for viral and microbial genomes) tracks representing genomic locations. A relatively small number of key tracks are shown by default when a genome is initially loaded, with more tracks loadable upon user request. Almost all genome browsers provide the user with the means to adjust the viewing scale, so that a whole genome can be viewed in its entirety, while specific genomic areas of interest can also be selected from the global view for more detailed viewing. The display content and layout can be customized to varying degrees. Most tools also have a built-in search capability to allow the user to search for specific annotations, such as gene names or gene families, with more complex queries of the underlying databases supported in certain genome browsers. As the number of Web-based tools and services for genomic data increases, some tools link to those external resources in a dynamic fashion, eliminating the need for reprogramming each time a new external resource needs to be accessed. Some genome browsers can also display multiple genomes simultaneously and show the similarities between annotated features for comparative genomics.

9.2 WEB-BASED GENOME ANNOTATION BROWSERS

Most Web-based genome browsers have originally been developed as a front end for accessing and viewing specific genome databases, which were created or maintained by the same group who developed the tool. Naturally, the development of this type of a genome browser is tightly coupled with the expansion of the genome databases that each site hosts. Nevertheless, some of these browsers over time developed into a modularized form that can be downloaded and installed to view locally hosted genome annotation data, which are stored using the same data format standard as the data for which the tool was initially created.

9.2.1 University of California, Santa Cruz (UCSC) Genome Browser

The University of California, Santa Cruz (UCSC) Genome Browser (Kent et al., 2002) was initially developed to make up-to-date versions of the human genome sequences available to the public as part of the Human Genome Project (www.genome.gov/11006939) (International Human Genome Sequencing Consortium, 2001). The UCSC browser now utilizes a MySQL-based database to catalog the genomic information but began earlier as a small script in the C programming language, which was initially intended to view splicing diagrams for gene prediction in the nematode *Caenorhabditis elegans* (Kamath et al., 2003). The move to use MySQL was borne out of the need to maintain program response times, while viewing significantly larger amounts of data in the human genome, the latest assembly of which is Genome Reference Consortium Human Build 37 (GRCh37, also known as hg19), released in February 2009.

Over the years, the UCSC Genome Browser database (Dreszer et al., 2012) has been expanded to provide access to genome annotations from many external sources as well as from UCSC. Updates on the database in terms of new genome assemblies and annotations are provided annually on the Database issue of the journal *Nucleic Acids Research* (http://nar.oxfordjournals.org). These sequences and annotations have been generated from a variety of sources, including the UCSC annotation pipeline and collaborators from research centers worldwide. UCSC manages the official repository of sequence-related data for the Encyclopedia of DNA Elements (ENCODE) consortium (http://genome.ucsc.edu/ENCODE) (ENCODE Consortium, 2004; Rosenbloom et al., 2012) and supports the coordination of data submission, storage, retrieval, and visualization. ENCODE started in September 2003 with the goal of identifying functional elements in the human genome sequence.

A unique feature of the UCSC browser is the main graphic representation of nucleotide sequences with the ability to load and display various annotations, referred to as "tracks." The many different tracks are separated into eight categories: mapping and sequencing, phenotype and disease associations, gene and gene prediction, messenger RNA (mRNA) and expressed sequence tag (EST), expression, regulation, comparative genomics, and variations and repeats. The UCSC browser allows for visualization of the many tracks available on the database. mRNA and EST data can be obtained from GenBank, with alignment to the genomic sequence performed by BLAT (Kent, 2002). The expression data can be obtained from collaborators, such as Affymetrix, or the Genomics Institute of the Novartis Research Foundation (GNF). Furthermore, variation information includes a presentation of loci from the Single Nucleotide Polymorphism Database (dbSNP; www.ncbi.nlm.nih.gov/projects/SNP). Repeats are presented from RepeatMasker (www.repeatmasker.org) and the Tandem Repeats Finder (TRF; http://tandem.bu.edu/trf/trf.html).

The UCSC browser user interface contains a list of drop-down menus that can be used to initiate the user session. The set of menu options allows users to focus on a particular aspect of an organism's genome that they may wish to examine. There is also a link that allows users to add custom tracks to the displayed data. Upon locating the proper region of interest, the Web site is redirected to a graphical view, displaying the region horizontally, along with a few tracks in that region. Menu options located at the bottom of this graphic display allow the user to customize annotations displayed on the screen. Zooming is performed by clicking on the scale buttons (1.5×, 3×, and 10×) that are displayed above the graphic. Additionally, the user can click on a specific position to jump to the base sequence of a specific segment of the chromosome. Moving left and right is also done by clicking on the arrow buttons adjacent to zoom and can also be done at three speeds. Figure 9.1 shows a screenshot of the UCSC browser display of a genomic region on human chromosome 21.

The UCSC browser has a variety of functions that can be used for research. These include the display of many different annotations, which are broken into several categories. In addition to these annotations, a homology search can be performed using BLAT, sequence searches using polymerase chain reaction (PCR) primer pairs with *in silico* PCR, and gene tables containing related genes based on homology, expression, or genomic location can be generated with Gene Sorter (http://genome.ucsc.edu/cgi-bin/hgNear). Also, users of the UCSC browser can upload their own annotations for display in the browser. Temporary additions can be accomplished in several formats including GFF, GTF, PSL, and BED file formats (see http://genome.ucsc.edu/FAQ/FAQformat for the descriptions of these formats) and can be viewed only on the machine from which the data was uploaded, for 8 hours after they were last accessed. Permanent and public additions can be done by uploading the track information onto a Web site and linking the Web address (i.e., the URL) to the genome browser.

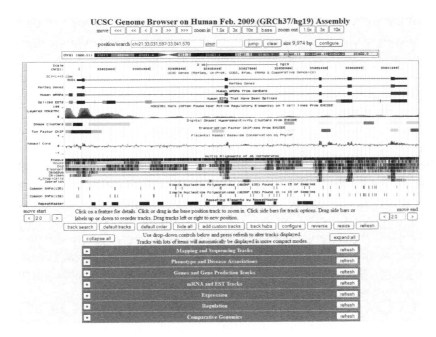

FIGURE 9.1 UCSC Genome Browser display of a human genome assembly. The top portion contains zoom scale and position control buttons, the middle portion is the main graphic showing various annotations, and the bottom portion has the menus for controlling display of tracks. **(See color insert.)**

9.2.2 Ensembl Genome Browser

The Ensembl Genome Browser (Flicek et al., 2011) is the main visualization tool for the Ensembl project, which generates databases for vertebrates and other eukaryotic genomes. As of Release 59 (August 2010), the Ensembl database hosts 56 species, most of which are vertebrates.

The Ensembl Genome Browser's individual species pages contain statistical information, including the known protein-coding genes, novel protein-coding genes, pseudogenes, microRNA (miRNA) genes, rRNA, small nuclear RNA (snRNA), small nucleolar RNA (snoRNA), and other RNA genes as well as single nucleotide polymorphisms (SNPs). RNA gene annotation in Ensembl is accomplished by making use of partial cDNA or EST and similarity to protein-coding genes found in other organisms. Non-coding RNA (ncRNA) determination is achieved through the collaboration with the RNA family database (Rfam) alignments of the Wellcome Trust Sanger Institute (www.sanger.ac.uk/Software/Rfam). In addition to the various RNA sequences annotated at Ensembl, resources are being dedicated to storing information regarding regulatory regions and variations. Information on variation is not limited to only SNP but also includes insertions and deletions. The development of this system is for storing both variations that are seen in natural populations and in lab-managed strains.

Ensembl utilizes different computational methods to create the various aspects of the program. The Ensembl Web site is written in Perl, and this code can be installed locally. The coding is modular and extensible so that the system may be customized to the demands of the researcher. The database is designed using MySQL relational databases. In addition, a comprehensive set of application programming interfaces (APIs) serve as a middle layer between underlying database schemes and more specific application programs, whose aims are to encapsulate database layout by providing efficient high-level access to data tables, and to isolate applications from data layout changes.

The Ensembl Genome Browser also uses a Web browser-embedded program to facilitate the visual illustration of genomes. In the opening page, the user can select a genome, which leads to various "entry points" into the genome annotation data, such as karyotype, location, gene, transcript, variation, phenotype, and regulation. For example, the karyotype entry point displays the organism's chromosomes, with summary information about the genes below. Clicking on the image of a chromosome gives the user the option of linking to either location view or chromosome summary. Figure 9.2 shows a location view, which contains a chromosome view and the overview of a selected region in the chromosome. The program also allows for jumping to a specific region of the genome after inputting a base pair position into the Location field. The width of the view can be adjusted with a drop-down menu. Furthermore, drop-down menus at the top allow the user to adjust the display of various data sets.

The Ensembl Genome Browser contains many functions that can be used to aid researchers. A variety of information can be displayed in the main graphic display, such as for restriction enzyme sites. In addition, genetic features, repetitive sequences, and decorations can also be added to the display. Genetic features include operons, RNA interference (RNAi) EST sequences, fosmids and gene predictions derived from other databases, EST information, and homologous proteins. Repeat sequences may include long terminal repeats (LTRs), RNA repeats, satellite repeats,

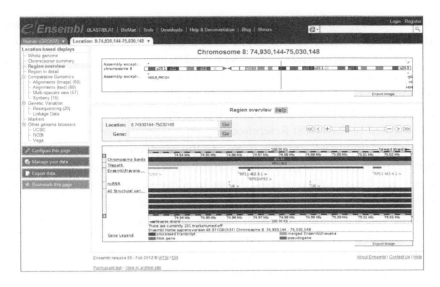

FIGURE 9.2   Ensembl Genome Browser display of Region Overview. The top portion (chromosome view) indicates the currently viewed region within the chromosome and the bottom portion shows the genomic features in several tracks. The bar in the middle allows the user to navigate to other regions or change the viewing scale. **(See color insert.)**

simple repeats, tandem repeats, long interspersed nuclear element (LINE) and short interspersed nuclear element (SINE) repeats, and Type II Transposons. Finally, decorations can also be added such as sequence information, codons, start and stop codons, rule and scale bars, G+C content, a gene legend or a SNP legend. Ensembl also provides the user with several features to customize the program. Individual data sets generated by the users can also be used and added into the Ensembl Genome Browser. Uploading of data can be achieved either directly onto the Ensembl server, to a public server and linking via the URL, or by setting up a Distributed Annotation Server (Dowell et al., 2001; Jenkinson et al., 2008).

9.2.3 NCBI Map Viewer

The NCBI Map Viewer (Wolfsberg, 2011) is a genome visualization tool that directly utilizes the vast amount of data and resources of the NCBI, including GenBank and PubMed. In particular, the Map Viewer provides special integrated maps for a subset of organisms available from Entrez Genomes (www.ncbi.nlm.nih.gov/sites/genome), which provides access to more than 800 complete genomes. The NCBI genome sequences are annotated to include coding regions, conserved domains, variation data, database cross-references, references, and names. This is accomplished using a combined approach of collaboration and external input from the scientific community, automated annotation, propagation from GenBank, and curation by the staff at NCBI. These sequences and annotations are represented in the NCBI Reference Sequence (RefSeq; www.ncbi.nlm.nih. gov/refseq) database, which is linked to the Map Viewer, UniGene (www. ncbi.nlm.nih.gov/refseq), HomoloGene (www.ncbi.nlm.nih.gov/homolo-gene), and UniSTS (Sayers et al., 2009). In addition to the RefSeq database, Repbase (www.girinst.org/repbase) contains a database of eukaryotic repetitive DNA elements, which include sequences of repeats and basic information described in the annotations.

The NCBI utilizes their own specialized software for the submission of data. RepbaseSubmitter (Kohany et al., 2006) is a Java-based interface for formatting and annotating entries into the Repbase database, which can be used to eliminate many common formatting errors. In addition, RepbaseSubmitter automates actions such as the calculation of sequence lengths and composition, which further facilitates the curation of Repbase sequences. The tool also has several features for the prediction of protein-coding regions in sequences, database searching, the inclusion of PubMed references in data sets, and searching of the NCBI taxonomy database for the correct inclusion of species-specific information and taxonomic position. To further augment this program, Censor (Kohany et al., 2006) is a tool that can rapidly identify repetitive elements by comparison to known repeats.

Map Viewer displays are provided at four levels of detail: (1) Home page, summary of the resources for the organism; (2) Genome View, display of the complete genome as scaled chromosome ideograms with search capability; (3) Map View, display of the regions of interest for a selected chromosome at various resolutions; and (4) Sequence View, display of the sequence data for a chromosomal region and the biological features of the region. The opening page (www.ncbi.nlm.nih.gov/mapview) has a list of organisms, with links to a genome view, which displays the chromosomes. Clicking on a chromosome will lead to a map view, an example of which is shown in Figure 9.3, which provides a general overview of the chromosome, including information on genes, description of the gene activity, and further links to other sources of information. In this view, users are able to drill down to detailed features of the chromosome, for example, to obtain information on the gene locations and sizes. In addition, a small

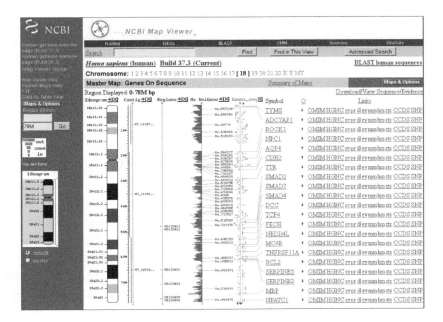

FIGURE 9.3   NCBI Map Viewer display of a Map View. The human chromosome 18 is shown vertically, along which linkable genomic features are shown. The display resolution can be changed by clicking on any vertical track and selecting either a zoom scale or a number of bases to show.

thumbnail of the entire chromosome is located on the left of the screen to indicate the currently displayed region within the chromosome and to allow for navigation throughout the chromosome.

The Map Viewer contains a number of interlinked services, which are connected to the genome sequences on the main page. Journal articles pertaining to a sequence element of interest, such as a gene, are directly linked to the graphical display, thus providing easy access to supplemental information. Furthermore, links are provided to the Online Mendelian Inheritance in Man (OMIM) resource (Sayers et al., 2009) and GenBank. This can provide further information regarding the gene, such as disease states, polymorphisms, and nucleotide sequence information. Also, a "Map and Options" button opens a window for customizing the display of certain information on the main genome graphic display. Users of the Map Viewer are able to create an account and upload data either from a local file or by inputting the URL of the file. Data can also be downloaded to the local machine in either Postscript image format for high resolution or as Adobe PDF files. However, the design of the Map Viewer does not allow it to be downloaded and customized to fit individual user needs, as its main purpose is to function as a viewing interface for the vast collection of genomic data hosted at the NCBI.

9.2.4 Generic Genome Browser

The Generic Genome Browser (GBrowse) (Stein et al., 2002) is one of the most widely used software tools among those developed from the Generic Model Organism Database Project (GMOD) (http://gmod. org). At the time of writing, according to the GMOD Users page (http:// gmod.org/wiki/GMOD_Users), there are more than one hundred databases or organizations that use GBrowse in one form or another. Some examples of the users are *Saccharomyces* Genome Database (SGD; http:// yeastgenome.org), WormBase (http://wormbase.org), FlyBase (http:// flybase.org), International HapMap Project (http://hapmap.org), Mouse Genome Informatics (MGI; http://informatics.jax.org), The *Arabidopsis* Information Resource (TAIR; http://arabidopsis.org), Gramene (http:// gramene.org), and Sea Urchin Genome Database (SpBase; http://spbase. org).

Genomic data is displayed at three different levels: overview, region, and details. At each level, annotation data are horizontally organized into tracks. Various genomic features can be easily represented on the tracks using a variety of premade and customizable glyphs, which are graphic symbols that differ in color, shape, and size to represent different objects. The vertical arrangement and appearance of the tracks can be customized via the user interface. Searches through the annotated features can be performed by annotation ID, keyword, name, or comment. Users can also upload custom tracks onto GBrowse, which includes third-party feature annotations. Arbitrary URLs can be attached to an annotation to link to external information resources. GBrowse also provides a session-management function through Web browser cookies, so that configuration settings can persist across sessions (as long as the cookies are not deleted, of course) and the browser resumes from where the user left off in the previous browsing session. Unlike the UCSC Genome Browser, the Ensembl Genome Browser, or the NCBI Map Viewer, GBrowse is not tied to particular genome databases; instead the system is designed to be portable and extensible. This allows the deployment of GBrowse as a customized genome browser interface for the use in the annotation of model organisms and other genome databases. GBrowse can be modified at the database layer, the data model layer, and the application layer. It is a downloadable, open-source package (http://sourceforge.net/projects/gmod/files) that has a flexible plug-in architecture, which allows developers to add new functionality without modifying the software source code. This is achieved by incorporating plug-ins through the plug-in API. The GBrowse distribution comes with some standard plugins, and installing custom plugins is as simple as copying a Perl module (.pm file) into a separate directory and setting a GBrowse variable to indicate the directory location where all of the plugins reside. Some examples of the use of plugins include running BLAST, dumping to and importing from many different formats, finding oligonucleotides, designing primers, and creating restriction maps.

9.3 STAND-ALONE GENOME ANNOTATION BROWSERS

Stand-alone genome browsers, as opposed to Web-based ones, are not necessarily closely tied to a specific genome database site but instead were mostly built as general-purpose browsers that can be downloaded and installed as an application on a local computer. These applications in general respond faster to user requests than Web-based browsers, since there is no need to communicate with a centralized server once the required genome annotation data is stored on the local disk. In addition, the ability to edit and save genome annotations is frequently available in these types of browsers. Most of these browsers are written in the Java programming language, as it enables the program to run without alterations on most computer platforms, and also operate through Java Web Start, which allows users to launch these programs from a Web browser without explicitly installing them in advance.

9.3.1 Bluejay: An XML-Based Genome Visualization and Data Integration System

The Bluejay genome browser (Soh et al., 2012; Soh et al., 2008) has been developed to meet the challenges posed by the increasing number of data types as well as the increasing volume of data generated through genome research. Bluejay started as a browser capable of rendering views of genomic information expressed in Extensible Markup Language (XML; www.w3.org/XML) and providing scalable vector graphics (SVG; www.w3.org/TR/SVG) output (Turinsky et al., 2004), which supports almost unlimited semantic zooming and publication-quality image output, but it now has many additional features. Recent developments focused on integrating gene expression profiling and comparative genomics into the browser to offer functional insights, which would have not been possible with traditional single-gene analyses and the display of their annotation results.

Bluejay is a genome viewer that is capable of integrating genome annotation with gene expression information. It also facilitates comparative genome analysis, based on gene paralogy information, with an unlimited number of other genomes in the same view. It allows the biologist to see a gene not just in the context of its genome but also its regulation and evolution (where gene expression data exist). Rich provision for personalization is available in Bluejay, including numerous display customization features and GPS-style waypoints for tagging multiple positions of interest within a genome for quick navigation to several genomic features. Bluejay also embeds the Seahawk browser (Gordon and Sensen, 2007) for the Moby protocol (BioMoby Consortium et al., 2008), enabling users to seamlessly invoke hundreds of Web services on various types of genomic data without any changes to the Bluejay code. Bluejay offers a unique set of customizable genome-browsing features, with the goal of allowing biologists to quickly focus on, analyze, compare, and retrieve related information on the parts of the genomic data in which they are most interested.

Genome sequence data first needs to be analyzed using an annotation pipeline, such as MAGPIE (Turinsky et al., 2005), in order to obtain meaningful visualization in Bluejay, as Bluejay requires a XML file with the genome annotation as the basis of the displayed figures. Minimally, the positions of the genes in a genome and their function call are required, in addition to the nucleotide sequence. Alternatively, annotated genomes, for example in GenBank XML format (for a full list of supported file formats, see below) available from public repositories (e.g., GenBank and any other resource that provides annotated genome files in a compatible format) can also be loaded into Bluejay. In either case, the resulting annotated sequence is loaded into Bluejay in either XML or non-XML format by requesting a URL or opening a local file. Bluejay has built-in bookmarks to load public genomes available from the MAGPIE home page (http://magpie.ucalgary. ca). Any non-XML data are converted internally to a Bioseq-compatible set of XML documents. Bluejay supports the following XML dialects for genome annotation:

- AGAVE (Architecture for Genomic Annotation, Visualization and Exchange), www.bluejay.ucalgary.ca/dtds/agave.dtd
- TIGR, the format in which TIGR (The Institute for Genomic Research) annotations are distributed, www.bluejay.ucalgary.ca/dtds/tigrxml.dtd
- Bioseq-set, the output format for XML option in readseq, a popular sequence format conversion program, www.bluejay.ucalgary.ca/dtds/Bioseq.dtd
- NCBI_GBSeq, NCBI GenBank's XML sequence data format, www.ncbi.nlm.nih.gov/dtd/NCBI_GBSeq.dtd

In addition to XML files containing genome annotation information, there are other sources of data for Bluejay. These additional data types can be visually integrated with the genome annotation. While the genome is displayed, the XLink standard (www.w3.org/TR/xlink) is used to hyperlink visual representations of genomic features to external sources of additional data for specific components of the genome. For example, MAGPIE gene annotation pages, if available, can be launched by clicking on a gene. In fact, custom XML documents created by users that describe a genome can include XLinks to any Web page. XLinks are inserted before the annotated sequence is internally represented as a Document Object Model (DOM; www.w3c.org/DOM) tree.

While using a genome viewer, biologists are often interested in visualizing only a small portion of a much larger genome. Navigating within a large genome to focus on the small region of interest usually involves many scrolling and zooming operations, until the desired part is in view at a proper zoom scale. This requires the user to go through several manipulations within the respective viewing environment to focus on the desired part. Another way is to type in the nucleotide position as a number to change the focus to center on that position. In this case, the user has to know or even memorize the position of the desired section within the genome. This is not very user friendly, and makes navigating among specific parts of a genome a tedious task.

Navigation within a genome is made significantly more convenient in Bluejay by introducing the concept of a waypoint, which is roughly defined as a coordinate that identifies a point in 2D or 3D space. In global positioning system (GPS) applications, waypoints correspond to coordinates of locations of interest on the earth's surface. Similarly, waypoints in Bluejay are a set of graphical flags or markers that are used to mark points of interest within the genome. Each waypoint contains the positional and other descriptive information (such as a potential function of the highlighted region) about that particular point. For example, the user can set a waypoint within a genome to highlight a specific gene or sequence feature, such as a transcriptional promoter or terminator. Figure 9.4 shows a typical use of waypoints, where the user first sets two waypoints and then uses the second waypoint to quickly focus on the second region of interest, while he can use the first one to jump back to the first region of interest at any time. This allows the rapid and efficient editing of attributes of these two genomic regions. Setting a waypoint automatically enables a set of operations, such as focusing on it, cutting the genome at it, and aligning several genomes at all waypoints labeled with the identical name, in multiple genomes.

Bluejay also allows the user to view a genome at various levels of detail, from the complete genome up to the nucleotide text level. In the text view mode, a waypoint can be set not only on a whole gene but also at any individual base position that is part of the genome. This helps the user to explore a collection of bases, even those that might be separated from each other by a large number of bases. For example, in the "horizontal sequential" text view mode of Bluejay, a genome is displayed as an elongated horizontal string of base characters.

Bluejay is written in Java and therefore can be run on any computer with a Java installation (version 1.5 or later). It is a free package downloadable from the Bluejay Web site (http://bluejay.ucalgary.ca). There are three options for download: (1) full application that can be installed on a local computer; (2) Java Web Start that can be run from a Web browser with Java plugin without local installation; and (3) Java applet for mainly testing purposes.

9.3.2 NCBI Genome Workbench

The NCBI Genome Workbench (www.ncbi.nlm.nih.gov/projects/gbench) is a customizable, integrated application for viewing and analyzing sequence data. It allows the users to organize and view data from publicly available NCBI sequence databases as well as their own private sequence data, or a mixture of both. The Genome Workbench offers several different ways of viewing sequence data, including graphical sequence views, multiple alignment and cross-alignment views, phylogenetic tree views, dot matrix views, and tabular views. Private data can also be aligned to

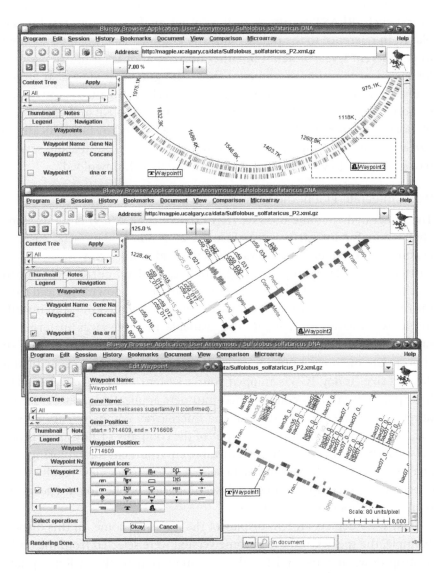

FIGURE 9.4 Customized navigation using waypoints in Bluejay. The user sets two waypoints on the displayed genome and zooms in on the area around "Waypoint2" (top). The selected region is zoomed in and displayed in more detail. Then the user clicks on "Waypoint1" in the "Waypoints" tab (middle). The focus changes to the region around "Waypoint1." Then the user selects "Edit Waypoint" operation to change the attributes of "Waypoint1" (bottom). **(See color insert.)**

public sequence data using the alignment tools provided in the Genome Workbench. BLAST results as well as other analysis results can be displayed in reference to the public data. The graphic display can be zoomed in at different levels of detail in either horizontal or vertical orientations for visual exploration of sequence data. The graphical view also allows users to select an object, such as a gene, and perform an analysis on it. The Genome Workbench is built using the NCBI C++ ToolKit and uses cross-platform APIs for graphics. It can be downloaded and installed locally to run as a stand-alone application on Windows, Mac OS, Linux, and several flavors of Unix (www.ncbi.nlm.nih.gov/tools/gbench/download).

9.3.3 Integrated Genome Browser

The Integrated Genome Browser (IGB) (Nicol et al., 2009) is stand-alone application for interactive visualization of genomic data sets, such as genome annotations, microarray designs, and tiling array data. IGB is written in the Java programming language and its current download mechanism uses Java Web Start, which automatically downloads and runs the required software without the user explicitly downloading and installing it (www.bioviz.org/igb). IGB is available in three sizes in terms of memory requirement (1 GB, 2 GB, and 5 GB), so that users can choose an appropriate memory footprint depending on their memory capacity and data set size. Today, IGB is an open-source system, allowing developers to incorporate it (and its components) into new applications, but its initial development was largely based at the microarray company Affymetrix, Inc., where it was used for the analysis of the company's tiling array products. IGB provides an easily customizable environment for exploring and analyzing large-scale genomic data sets. IGB lets users view results from experiments or computational analyses alongside public domain genomic sequence, annotations, and data sets.

Figure 9.5 depicts an IGB display of the mouse genome. In this display example, the species and the genome version were selected from the built-in drop-down menus, and the default annotation tracks of RefSeq data for the selected chromosomal region is shown. Data can be loaded into IGB from several sources (using the items in the collapsible list in the Data Access tab) including from the Distributed Annotation Servers (Dowell et al., 2001), the QuickLoad server (http://igbquickload.org/das2), any Web site containing links to the appropriate data, and local files. Web control is possible with IGB such that it can receive HTTP requests from any program or a Web browser in order to view specific regions of local data sets in

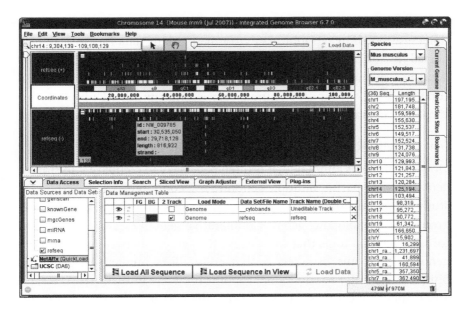

FIGURE 9.5   Integrated Genome Browser (IGB) display of mouse chromosome 14. When a genome is loaded using the built-in Species and Genome Version drop-downs, the RefSeq data tracks are displayed by default. Genomic data from other sources (e.g., QuickLoad and DAS) can be loaded within the Data Access tab.

a Web page using IGB. Users can also write simple scripts to direct IGB to load a genome, zoom and change focus to specific regions, and show associated data. Other features of IGB include a special zooming display, where successive resolutions are shown when the zoom scale slider is moved. This works all the way up to individual nucleotides, avoiding jumping between zoom scales, and allowing users to easily find the most appropriate zoom scale.

9.3.4 Apollo

Apollo (Lewis et al., 2002) is a genome annotation editing and viewing tool. It was jointly developed by the Berkeley *Drosophila* Genome Project (http://apollo.berkeleybop.org), which is part of the FlyBase consortium and the Sanger Institute (http://ensembl.org) of Cambridge, United Kingdom. It was initially used in FlyBase to annotate the *Drosophila melanogaster* genome (Celniker and Rubin, 2003). Apollo has been incorporated in the GMOD project as the main community annotation and sharing tool through which data providers can share their annotations and let collaborators edit them directly. Apollo allows researchers to visualize genomic annotations at different levels of detail and to perform annotation curation all in a graphical environment. Many data formats are supported in Apollo, including XML, GenBank, GFF3, and GMOD's Chado database schema. Apollo is a Java application that can be downloaded and run on Windows, Mac OS X, Linux, Solaris, and any other Unix platforms capable of operating Java.

9.4 COMPARATIVE VISUALIZATION OF GENOMES

Once a genome is annotated, it is often necessary to compare it to similar genomes. Comparing multiple genomes is a commonly used method for studying organisms, to elucidate the functions and evolutionary background of homologous genes. Comparative genomics created the need for visualization tools that can display the synteny between two or more genomes. Genome browsers with facilities for comparative genomics built in are similar to traditional genome browsers in their user interface and display appearance, but have generally two additional key capabilities built in: a display feature for more than one genome in the display area, and some means to visually express the relationship between the displayed genomes or genomic features. Genome browsers with comparative features can be roughly classified into three groups, according to how relationships between compared genomes are visually depicted: tools capable of displaying dot plots, linear representations, or circular representations.

9.4.1 Dot Plots

Two-dimensional dot plots have traditionally been used to visualize the relationship between two data sets, where one set is represented along the horizontal axis and the other set along the vertical axis. The horizontal and vertical axes each represent a different genome or species, with the dots in the plot representing similar genomic elements. This display scheme is thus useful for showing the alignment of only two genomes, which can be a severe limitation. It is implemented in tools such as SynMap (Lyons et al., 2008), VISTA (http://pipeline.lbl.gov), MEDEA (www.broadinstitute.org/annotation/medea), and GenomeMatcher (Ohtsubo et al., 2008). Gene duplication or rearrangement can be easily identified by local groups of dots, where each group forms a certain clustered pattern such as a straight line. Zooming and panning are usually possible with these tools, allowing the user to navigate through different parts of the displayed genomes.

9.4.2 Linear Representation

Traditional Web-based genome browsers usually use the concept of a track to show a specific type of a genomic feature at single or multiple genomic locations. As a consequence, the vast majority of comparative genome browsers also rely on multiple linear tracks to represent multiple genomes and their features. Three genome browsers based on GBrowse alone have been developed to visualize genome comparison. Being based on GBrowse, all of these browsers display additional genomes for comparison in a linear track representation. GBrowse_syn (http://gmod.org/wiki/GBrowse_syn), or the Generic Synteny Browser, can be used to display multiple genomes together with a reference genome. The reference genome is displayed at the center, with additional genomes displayed above and below. Multiple sequence alignment, synteny, or colinearity data from other sources can be viewed against GBrowse-produced genome annotations. Users can change the central genome (i.e., the reference genome) as well as the unit size for comparison display, meaning that genomes can be compared at the sequence level, gene level, or larger block level. The current standard GBrowse package includes GBrowse_syn tool kit.

SynBrowse (Brendel et al., 2007) is a software tool within the GBrowse family for generic sequence comparison. This is achieved by visualizing genome alignments between different species and also within the same species, if desired. The design goal of SynBrowse is to help biologists analyze synteny, gene homology, and other kinds of conservation across genomes, which makes it useful for studying gene duplication and the evolution of genomes. Data for SynBrowse are created in standard GFF2 format (http://gmod.org/wiki/GFF2), where target tags are used to denote the correspondence of a region in a genome to another region in a different genome. The relationship between genomic features can be displayed at multiple levels, including "synteny block," "coding gene," and "coding exon." SynBrowse runs on top of the BioPerl (Stajich et al., 2002) modules and also uses the configuration and utility modules of GBrowse.

SynView (Wang et al., 2006) is another comparative genome browser that is made possible by the use of the standard GBrowse distribution and a sophisticated standard GBrowse configuration file. Perl callbacks are used to draw a comparison view on the GBrowse details panel, based on the synteny data stored in standard GFF3 format. SynView is included in the GBrowse distribution. SynView is used in PlasmoDB (www.plasmodb.org).

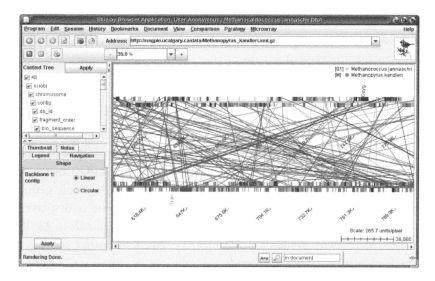

FIGURE 9.6 Comparison of two archaeal genomes in Bluejay in linear representation. Comparing *Methanococcus jannaschii* and *Methanopyrus kandleri* shows that they share many genes in the same Gene Ontology (GO) categories, indicated by the linking lines. **(See color insert.)**

In addition to the GBrowse family, there are a number of comparative genome browsers that have a built-in capability for the linear representation of genomes. These include Bluejay (Soh et al., 2012; Soh et al., 2008), Cinteny (Sinha and Meller, 2007), Apollo, MEDEA (www.broadinstitute. org/annotation/medea), Sybil (http://sybil.sourceforge.net), PhIGs (Dehal and Boore, 2006), VISTA (Mayor et al., 2000), VISTA Synteny Viewer (http://genome.jgi-psf.org/synteny), UCSC Genome Browser, CMap (Youens-Clark et al., 2009), ACT (Carver et al., 2008), Combo (Engels et al., 2006), and MultiPipMaker (Elnitski et al., 2005). Figure 9.6 shows the comparison of two archaeal genomes in Bluejay, where the two genomes are represented as two parallel linear stretches and two genes are linked if they share the same Gene Ontology (GO) classification. The comparative genomics functionality of Bluejay can be repurposed for comparing multiple chromosomes in a single genome. An example is shown in Figure 9.7, where the ability of Bluejay to compare more than two genomes is applied to the comparison of four human chromosomes or chromosomal arms, by treating the chromosomes as if they were genomes for visualization purposes. It has long been observed that human chromosomes contain many instances of gene family duplications (Dehal and Boore, 2005). Bluejay

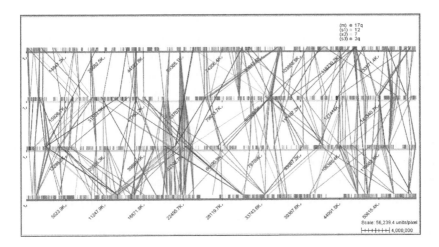

FIGURE 9.7   Comparison of multiple chromosomes in Bluejay. Human chromosomes 17q, 12, 7, and 2q are compared in Bluejay, which shows that many genes are duplicated on several chromosomes, as represented by the many lines that link genes belonging to the same gene family. (**See color insert.**)

allows the visual determination of whether a particular subset of human chromosomes indeed contains a set of duplicated genes. In Bluejay, the direction of gene linking is by default from the first genome to the second genome to the third genome, and so on, but this direction can be reversed, or even both directions can be considered. For example, in the comparison snapshot of Figure 9.7 chromosome 17q is the first genome, and the links are generated not only in the direction of bottom-to-top chromosome loading order (17q, 12, 7, 2q) but also in the top-to-bottom direction.

9.4.3 Circular Representation

The circular representation of genomes facilitates the intuitive comparison of multiple microbial or viral genomes in a whole-genome view. In this viewing mode, linear tracks are replaced by concentric circles, and arcs are used to represent genomic sections. Despite the obvious advantage, there are only a few genome browsers that facilitate this type of representation. In Bluejay, users can visualize multiple genomes side by side in a single display for direct visual comparison based on the genes rather than the nucleotides. Bluejay enters and exits the comparison mode according to the number of genomes currently loaded and allows a virtually unlimited number of genomes to be compared simultaneously.

The key feature of the comparison mode that facilitates the visual comparison is the display of lines that link common genes based on a consistent gene classification scheme. For example, Bluejay has a default bookmark that links to the genome data already annotated using the MAGPIE system, which uses GO (www.geneontology.org) as the main gene classification system. The GO project provides a controlled vocabulary to hierarchically describe gene and gene product attributes in any organism. Thus, when these genomes are loaded for comparison, each gene is linked to the closest gene with the same GO classification (close in terms of coordinates, not gene function) at the most detailed classification level. Using this feature, the user can quickly determine the degree of similarity between the genomes being compared, in terms of the overall distributions of genes within a genome (i.e., synteny). It is important to note that classification by GO categories is only one example of the possible gene classification systems that might be used in Bluejay for the linking of genes. Any sort of gene classification system that fits the purpose of genome comparison and provide a textual description of the defined gene categories can be used without any modifications to Bluejay, as long as it is included in the XML file that contains the genome annotation.

The visual comparison of two or more similar genomes at the gene level is often quite complicated, as groups of genes are often located in different parts of the genomes due to insertion, deletion, duplication, or translocation events during the genome evolution. Genomes can automatically be aligned in Bluejay to minimize the influence of genomic positional differences in the visual comparison. This is possible because in the comparison mode of Bluejay; all genome displays are scaled with the first loaded genome as the reference. As a consequence, all circular genomes are displayed as a complete circle and all linear genomes are of the same displayed length. Thus, an optimal alignment that minimizes the influence of gene rearrangements on comparison is determined by minimizing the sum of the angular distances for all linked gene pairs in the case of circular genomes. By rotating the outer sequence to the best-aligned position with respect to the inner sequence, this feature allows the user to visualize the best global alignment of closely related genes. The user can then see the functional similarity of the genes in the two genomes, with the effect of purely positional differences minimized as much as possible.

Another feature to aid gene category linking is the ability to selectively show or hide the links between genes by applying a threshold value on the linking distance. For linearly shaped genomes, a threshold

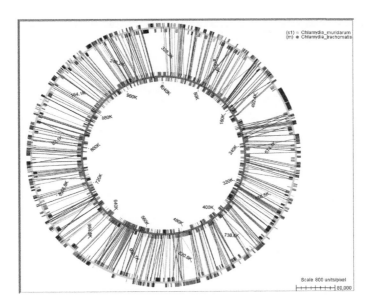

FIGURE 9.8 Comparison of two bacterial genomes in Bluejay in circular representation. Comparing *Chlamydia trachomatis* and *Chlamydia muridarum* reveals that they have many common genes according to the Gene Ontology (GO) classification, as shown by the linking lines. Only those links with an angular distance less than 2% of the maximum possible distance (360 degrees) are shown. **(See color insert.)**

value is set as a percentage of the length of the first genome loaded, whereas for circularly shaped genomes it is set as a percentage of 360°. By experimenting with this adjustable threshold value, the user can visually perceive the degree of similarity of the compared genomes by observing how many of the links are shorter than a certain distance in genomic position. Figure 9.8 shows an example of genome comparison in Bluejay, where two circular bacterial genomes are displayed together with genes of the same GO classification linked, with the linking distance threshold applied.

The availability of waypoints can augment the genome comparison functionality of Bluejay. It is often necessary to align multiple genomes at specific gene locations to investigate the similarity of the genomic region around those genes. The waypoint features in Bluejay provide the user with the ability to flag multiple genes, give the waypoints the same name in multiple genomes, and subsequently align the genomes at the respective flags, eliminating the need to estimate the amount of required rotation before the

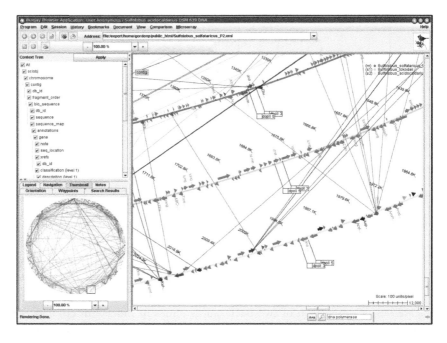

FIGURE 9.9 Genome alignment by waypoints in Bluejay. A waypoint named "dpoII" is set at the appropriate location in each of three *Sulfolobus* spp. genomes. On selecting "Align at Waypoint" operation, the genomes are aligned at the dpoII gene by appropriately rotating the two outer genomes. This facilitates easy visual comparison of the gene's structure in all three species. **(See color insert.)**

alignment. Without these Bluejay features, the user would laboriously try to rotate the genomes by some angles until the genes of interest align exactly. Figure 9.9 shows an example of aligning three genomes by a waypoint representing the gene of interest. This usage of waypoints in combination with automatic genome rotation for genome comparison adds greatly to the utility of both waypoints and genome comparison capability in Bluejay.

Circos (Krzywinski et al., 2009) is a software package for visualizing genomic information and the relationship between genes in a circular fashion. Like Bluejay, it uses layers of circles or arcs to express the positional layout of multiple genomes. Links are created between circular positions to express their relationship. The visual appearance and layout can be customized through the modification of plain-text configuration files. This lack of graphical user interface makes it more difficult to learn the use of this tool than display customization controlled by graphical user interface (such as the one in Bluejay) would allow. On the other hand, it is easier to incorporate the automated image creation step into data processing pipelines. Other tools for circular visual comparison of genomes include MEDEA (www.broadinstitute.org/annotation/medea) and MizBee (Meyer et al., 2009).

REFERENCES

BioMoby Consortium, Wilkinson, M.D., Senger, M. 2008. Interoperability with Moby 1.0—It's better than sharing your toothbrush! *Brief. Bioinformatics* 9(3):220–231.

Brendel, V., Kurtz, S., Pan, X. 2007. Visualization of syntenic relationships with SynBrowse. *Methods Mol. Biol.* 396:153–163.

Carver, T., Berriman, M., Tivey, A., et al. 2008. Artemis and ACT: Viewing, annotating and comparing sequences stored in a relational database. *Bioinformatics* 24:2672–2676.

Celniker, S.E., Rubin, G.M. 2003. The *Drosophila melanogaster* genome. *Annu. Rev. Genomics Hum. Genet.* 4:89–117.

Dehal, P., Boore, J.L. 2005. Two rounds of whole genome duplication in the ancestral vertebrate. *PLoS Biol.* 3(10):e314.

Dehal, P.S., Boore, J.L. 2006. A phylogenomic gene cluster resource: The Phylogenetically Inferred Groups (PhIGs) database. *BMC Bioinformatics* 7:201.

Dowell, R.D., Jokerst, R.M., Day, A., Eddy, S.R., Stein, L. 2001. The distributed annotation system. *BMC Bioinformatics* 2:7.

Dreszer, T.R., Karolchik, D., Zweig, A.S., et al. 2012. The UCSC Genome Browser database: Extensions and updates 2011. *Nucleic Acids Res.* 40(Database issue):D918–D923.

Elnitski, L., Riemer, C., Burhans, R., Hardison, R., Miller, W. 2005. MultiPipMaker: Comparative alignment server for multiple DNA sequences. *Curr. Protoc. Bioinformatics* Chapter 10:Unit 10.14.

ENCODE Consortium. 2004. The ENCODE (ENCyclopedia Of DNA Elements) Project. *Science* 306(5696):636–640.

Engels, R., Yu, T., Burge, C., et al. 2006. Combo: A whole genome comparative browser. *Bioinformatics* 22:1782–1783.

Flicek, P., Amode, M.R., Barrell, D., et al. 2011. Ensembl 2011. *Nucleic Acids Res.* 39(Database issue):D800–806.

Gordon, P.M.K., Sensen, C.W. 2007. Seahawk: Moving beyond HTML in Web-based bioinformatics analysis. *BMC Bioinformatics* 8:208.

International Human Genome Sequencing Consortium. 2001. Initial sequencing and analysis of the human genome. *Nature* 409:860–921.

Jenkinson, A.M., Albrecht, M., Birney, E., et al. 2008. Integrating biological data—The Distributed Annotation System. *BMC Bioinformatics* 9(Suppl 8):S3.

Kamath, R.S., Fraser, A.G., Dong, Y., et al. 2003. Systematic functional analysis of the *Caenorhabditis elegans* genome using RNAi. *Nature* 421(6920):231–237.

Kent, W.J. 2002. BLAT—The BLAST-like alignment tool. *Genome Res.* 12(4):656–664.

Kent, W.J., Sugnet, C.W., Furey, T.S., et al. 2002. The human genome browser at UCSC. *Genome Res.* 12:996–1006.

Kohany, O., Gentles, A.J., Hankus, L., Jurka, J. 2006. Annotation, submission and screening of repetitive elements in Repbase: RepbaseSubmitter and Censor. *BMC Bioinformatics* 7:474.

Krzywinski, M., Schein, J., Birol, I., et al. 2009. Circos: An information aesthetic for comparative genomics. *Genome Res.* 19:1639–1645.

Lewis, S.E., Searle, S.M.J., Harris, N., et al. 2002. Apollo: A sequence annotation editor. *Genome Biol.* 3(12):research0082.

Lyons, E., Pedersen, B., Kane, J., et al. 2008. Finding and comparing syntenic regions among *Arabidopsis* and the outgroups papaya, poplar, and grape: CoGe with rosids. *Plant Physiol.* 148:1772–1781.

Mayor, C., Brudno, M., Schwartz, J.R., et al. 2000. VISTA: Visualizing global DNA sequence alignments of arbitrary length. *Bioinformatics* 16:1046–1047.

Meyer, M., Munzner, T., Pfister, H. 2009. MizBee: A multiscale synteny browser. *IEEE Trans. Vis. Comput. Graph.* 15:897–904.

Nicol, J.W., Helt, G.A., Blanchard, S.G. Jr., Raja, A., Loraine, A.E. 2009. The Integrated Genome Browser: Free software for distribution and exploration of genome-scale data sets. *Bioinformatics* 25:2730–2731.

Ohtsubo, Y., Ikeda-Ohtsubo, W., Nagata, Y., Tsuda, M. 2008. GenomeMatcher: A graphical user interface for DNA sequence comparison. *BMC Bioinformatics* 9:376.

Rosenbloom, K.R., Dreszer, T.R., Long, J.C., et al. 2012. ENCODE whole-genome data in the UCSC Genome Browser: Update 2012. *Nucleic Acids Res.* 40(Database issue):D912–D917.

Sayers, E.W., Barret, T., Benson, D.A., et al. 2009. Database resources of the National Center for Biotechnology Information. *Nucleic Acids Res.* 37(Database issue):D5–D15.

Sinha, A.U., Meller, J. 2007. Cinteny: Flexible analysis and visualization of synteny and genome rearrangements in multiple organisms. *BMC Bioinformatics* 8:82.

Soh, J., Gordon, P.M.K., Sensen, C.W. 2012. The Bluejay genome browser. *Curr. Protoc. Bioinformatics* 10.9.1–10.9.23.

Soh, J., Gordon, P.M.K., Taschuk, M.L., et al. 2008. Bluejay 1.0: Genome browsing and comparison with rich customization provision and dynamic resource linking. *BMC Bioinformatics* 9:450.

Stajich, J.E., Block, D., Boulez, K., et al. 2002. The Bioperl toolkit: Perl modules for the life sciences. *Genome Res.* 12(10):1611–1618.

Stein, L.D., Mungall, C., Shu, S., et al. 2002. The generic genome browser: A building block for a model organism system database. *Genome Res.* 12(10):1599–1610.

Turinsky, A.L., Ah-Seng, A.C., Gordon, P.M.K., et al. 2004. Bioinformatics visualization and integration with open standards: The Bluejay genomic browser. *In Silico Biology* 5(2):187–198.

Turinsky, A.L., Gordon, P.M.K., Xu, E.W., Stromer, J.N., Sensen, C.W. 2005. Genomic data representation through images: The MAGPIE/Bluejay system. In *Handbook of Genome Research*, ed. C.W. Sensen, 187–198. Weinheim: Wiley-VCH.

Wang, H., Su, Y., Mackey, A.J., Kraemer, E.T., Kissinger, J.C. 2006. SynView: A GBrowse-compatible approach to visualizing comparative genome data. *Bioinformatics* 22:2308–2309.

Wolfsberg, T.G. 2011. Using the NCBI Map Viewer to browse genomic sequence data. *Curr. Protoc. Hum. Genet.* Chapter 18:Unit 18.5.

Youens-Clark, K., Faga, B., Yap, I.V., Stein, L., Ware, D. 2009. CMap 1.01: A comparative mapping application for the Internet. *Bioinformatics* 25:3040–3042.

CHAPTER 10

Web-Based Workflows

10.1 INTRODUCTION

The annotation pipelines discussed in Chapter 8 have fairly fixed inputs, analysis steps, and outputs. Dynamic display systems such as those discussed in Chapter 9 can help achieve an individually customized view of the annotation results. On the other hand, redefining the annotation outputs in existing pipelines is very difficult for a nonprogrammer. In this chapter, we explore the means by which a user with limited to no programming abilities can create their own analysis pipelines, for example, to supplement standard genomic pipelines with additional evidence or to find a subset of genes of interest. In general terms, the principles behind end-user pipeline development tools (also known as workflow) are introduced. Scientific workflow editors for local resources, such as Kepler (Ludaescher et al., 2006), will not be explored further here, as they do not play a major role in the field of genome annotation at this point in time. Three Web-based pipeline tools, which do not require a large local informatics infrastructure, are discussed in detail in this chapter.

10.2 PRINCIPLES OF WEB-BASED WORKFLOWS

10.2.1 Motivation

Besides the obvious reason that automation provides efficiency for custom analysis of genome-scale data sets, there are several reasons for adopting workflows in bioinformatics. First, a formal workflow can be used to document the analysis process used to derive the annotation results. This document often provides useful information for the "Methods" section when the results are eventually published. There is a growing trend to include formal workflows as supplementary information for papers, because they are communicative and provide the background for the reproducibility of results, which can be helpful for readers and reviewers alike. Workflows can be rerun periodically to ensure that the genome annotation results are up to date, with respect to the reference databases they were compared to, and by extension the current scientific knowledge base. Creating a workflow can save effort down the road through its reuse. For example, through slight modifications it might be used to perform different but related analyses in follow-up research.

Web-based workflows may take different forms, mainly either as workflow environments that use local analysis tools but are accessed via a Web portal interface, or as desktop workflow environments that coordinate Web-based analysis services. Both types of Web-based workflows are described in some detail, with Galaxy in the first category, and Taverna and Seahawk both being in the latter.

10.2.2 Early Workflow Environments

In 2004, de Knikker et al. (2004) provided a bioinformatics analysis scenario involving Web services. They tried to implement this in three ways: with a Java program, with a business workflow tool, and using Taverna (Oinn et al., 2004). They concluded that the Java program was the most straightforward approach, but this conclusion was derived from the perspective of the professional programmer. The programming of scientific workflows has seen a gradual evolution toward graphical editors and workflow component repositories. Graphical editing tools simplify the process of custom programmatic analysis by biologists, as they require less local resources and less knowledge of a programming syntax. A few prominent early tools for bioinformatics workflow automation are:

- BioPipe (Hoon et al., 2003) uses a hand-written XML specification file for the workflow and is implemented as a set of command-line Perl scripts interfacing to the BioPerl (Stajich et al., 2002) library of bioinformatics analysis modules.

- Pegasys (Shah et al., 2004) has a graphical workflow builder for accessing locally installed tools. The system is backed by a custom database.

- Integrator (Chagoyen et al., 2004) uses XQuery (www.w3.org/XML/Query) for dissecting and coordinating disparate Web-based resources. It also includes a custom procedural instruction set to integrate data from multiple Web resources using collection and filtering steps. It relies on the pertinent data being available in a data format known as XML.

- Wildfire (Tang et al., 2005) includes a graphical workflow builder with more advanced control elements, such as for-each loops. For the analysis steps, it employs grid-based computing for large data sets, with programs accessed via command-line interface definitions in EMBOSS format (Rice et al., 2000).

- JOpera (Pautasso and Alonso, 2005) started as a bioinformatics workflow tool but became more business-oriented during its development. It includes spreadsheet-like evaluation functions, integrates the Eclipse editor (www.eclipse.org) and uses Web Services (see Section 10.4) for the analysis tasks.

- Bio-STEER (Lee et al., 2007) has a relatively simple graphical interface and workflow composition mechanism, and uses World Wide Web Consortium's OWL-S standard (www.w3.org/Submission/OWL-S) to describe Web services, with knowledge domain-specific labels.

Three more contemporary approaches to workflow automation are described in more detail next.

10.3 GALAXY

10.3.1 Interactive Analysis

The Galaxy Web site (www.usegalaxy.org) provides a Web portal for running large-scale bioinformatics analyses (Goecks et al., 2010). The typical Galaxy user interacts with the system stepwise, interactively analyzing data files they upload to the system. Entries in files containing multiple records are all automatically processed the same way, allowing users to process large-volume data sets efficiently. Figure 10.1 shows the Web interface for this interactive analysis process. The interface has three main vertical panes, with the list of available actions on the left, the action input form in the middle, and the action history of the current session on the right. The first step in the analysis is typically to upload a data file or to load data via a query to an online database, such as Ensembl (Flicek et al., 2010). If the user uploads a data file, Galaxy will attempt to determine the

FIGURE 10.1   Galaxy interface for uploading data files. The file type uploaded in the selection box (a BAM file, second input in center pane) and the reference genome selected (hg19, last input in center pane) constrain what subsequent analyses can be run.

file format automatically, and thus the nature of its contents. Galaxy uses a "walled-garden" approach to analysis; a specific set of file formats are accepted, and analysis options can only be run with compatible files. These restrictions ensure that the analysis will always proceed smoothly, with the trade-off of making the integration of new tools and data types somewhat onerous (see Section 10.3.3), especially for nonprogrammers.

Specifically in a genomics context, uploaded files are often implicitly tied to a particular genomic reference sequence. As an example, when a BAM short-read alignment file is generated by comparing next-generation sequencing outputs (ABI SOLiD or Illumina) to a longer reference sequence, Galaxy requests this information ("Genome" at the bottom of Figure 10.1) in order to provide more appropriate options for the visualization and database cross-linking in subsequent steps.

Once an upload is complete, the corresponding history item in the right panel turns green. If a Galaxy tool (left panel) is selected and accepts the uploaded file type, a tool run configuration form is displayed, as shown in the center pane of Figure 10.2.

If the tool was not compatible with the uploaded data, the second drop-down box (BAM file) would be empty, and hence the tool ("pileup" in this example) could not be executed. After pileup is run, the user can employ other tools, which take a BAM file as input, and tools that can take pileup-generated files as output. By default, the results of an analysis are not displayed. To visualize particular results, the user must press the eye icon in the upper right hand corner of the history pane of the desired step.

FIGURE 10.2 Galaxy interface for running an analysis tool. The BAM file is automatically populated as a tool input (second input field in center pane) if it has finished loading (box in right pane).

FIGURE 10.3 Galaxy interface for running another analysis tool. (Center) Galaxy tool input with input auto-populated from the analysis run in Figure 10.2. (Right) Galaxy options for the analysis history, including extracting the three analysis steps as a workflow.

10.3.2 Workflows

The History pane accumulates a list of the tools that have already been used and which outputs were used as inputs for other tools, respectively. This history can be revised using the pencil and X icons, which are displayed in the upper right hand corner of each recorded history step. The analysis history can also be exported as a workflow by selecting the "Extract work-flow" item from the History pane's "Options" drop-down box, as shown in Figure 10.3. This constitutes what is called workflow-by-example functionality, where a workflow is generated without the user *explicitly* performing any programming; the program is *inferred* by the Galaxy system rather than written by a user and then added to Galaxy.

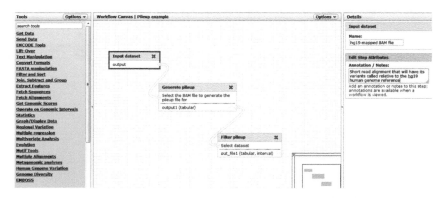

FIGURE 10.4   Galaxy graphical editor view of the workflow extracted in Figure 10.3. New tools can be added from the left pane and existing steps/connections edited on the diagram. Labels can be edited in the right pane.

The workflow, which was saved by the user in Galaxy, can be executed or edited graphically as in Figure 10.4. It is recommended to edit the workflow to include meaningful names for the workflow steps and free text annotations (right side of the figure), which can contain additional documentation. This maximizes the usefulness of the workflow, when revisited later, or by users with whom it is subsequently shared. The workflow can be edited later, for example, to reflect changing analysis needs. The layout of the steps, as well as the tool connection during the dataflow, can be directly manipulated in the workflow diagram. Workflow elements can be removed from the diagram and new ones can be added using the tool selection pane in the left-hand pane of the Galaxy interface (Figure 10.4).

It should be noted that running sequence-based workflows on the public Galaxy server can quickly exceed the allotted disk quotas. Therefore, workflows can be exported and executed on a local installation of Galaxy, but one must ensure that the local installation has the same set of tools enabled as the public portal.

10.3.3 Component Repository

Creating a local installation of Galaxy requires at a minimum a level of comfort with the command line, a multi-CPU computer, and significant available disk space to store reference databases and user analyses. Beyond avoiding quota limitations, a local installation allows the technically savvy user to deploy new file types and tools in the walled garden. The Galaxy Tool Shed (http://toolshed.g2.bx.psu.edu) is a repository of user-contributed noncore extensions to Galaxy. If maintained properly, a local Galaxy installation can stay better abreast of recent developments in sequence analysis techniques by using Shed tools, rather than relying on updates to the main public portal.

10.4 TAVERNA

Taverna (Oinn et al., 2004) is a popular workflow editing and execution environment that focuses on choreographing Web-based resources, as opposed to Galaxy's ability to use a fixed set of local analysis tools. The programming of Taverna workflows is done by explicitly building a graphical workflow diagram. In particular, Taverna focuses on a particular set of Web-based resources, called Web services, which are formally described and easily accessed programmatically. At this point in time, most popular bioinformatics sites also provide access to a Web Services mode of their analysis tools. This avoids the user having to write "screen scraper" code to parse the human-readable versions of the information served by the site. A registry of bioinformatics-related Web services is maintained on the Web site BioCatalogue.org (Bhagat et al., 2010).

A Taverna Web service workflow consists of four main component types:

- Workflow inputs—This is typically a list of data of some sort, such as a spreadsheet of values (e.g., text file or Microsoft Excel file) or manually recorded values.

- Processors—These are the functions that transform data, such as running BLAST, or tools for removing duplicate data. A processor can be Web service, a BioMart query, a BeanShell script (Java code snippet), an R script, or a call to a local command-line tool. Constant values are a type of processor too, usually shown in blue. A processor type that was recently added to Taverna is the REST API (Fielding and Taylor, 2002), which can be used to retrieve data by calling a carefully crafted Web address. For example, a particular Web address template can be used to retrieve a text-formatted database entry from UniProt, when given the UniProt ID (www.uniprot.org/uniprot/UNIPROTID.txt, where UNIPROTID is replaced with a real UniProt ID, e.g., P68431).

- Data links—These are arrows that direct the flow of data from the output of one processor to become the input of another.

- Workflow outputs—These are the results that users are interested in obtaining. Intermediate results from each stage of the workflow can be listed as well, but require more work to access.

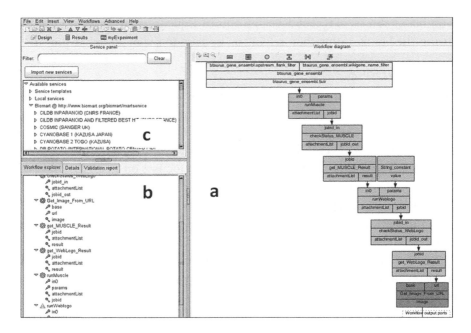

FIGURE 10.5   Taverna interface in Design mode. Main components are: (a) interactive graphical workflow diagram; (b) workflow explorer, hierarchical listing of workflow inputs, workflow outputs, processors with their ports, and data links between ports; and (c) list of available processors for inclusion in the workflow. **(See color insert.)**

The two main views of the workflow are Design (editing mode) and Results (execution mode). Both of these are explained next.

10.4.1 The Design Interface

Figure 10.5 shows the workflow Design interface for Taverna. This is the interface that is displayed when an existing workflow is first loaded or a new one is created. There are three main parts to the Design interface in Taverna (see the panes labeled a, b, and c in Figure 10.5):

    a. The workflow diagram—A visual depiction of the flow of data between processors. Built-in processors are colored in purple and Web services are colored in green. It is recommended that the view that includes all of the input and output ports (fifth icon from the left at the top of the diagram pane) is used to fully understand the connection between the various processors.

b. The workflow explorer—Each component of the workflow is listed, including the input and output parameters or ports for each processor. If a component is selected, the content of the "Details" tab in the same pane is changed correspondingly. Users can flip between these tabs by clicking their labels at the top of the pane.

c. The processors list—A list of the tools that could be added to the workflow. This list is searchable using the "Filter widget" directly above the list.

In terms of user interactions, left-clicking selects interface components, as expected. Right-clicking reveals (for most components) the options that are available, either in the workflow diagram, the workflow component list, or the processor list. These options may include, for example, the renaming of a component, the setting of a parameter value, or the linking of processors. Left-button dragging from an output port to an input port in the workflow diagram also creates a data link between the two elements. Although almost any port can be linked together with another one, not every combination of ports will yield a reasonable result. For example, if the output of a service is a DNA sequence, but the next stage of the analysis takes an NCBI identifier as input, Taverna will allow the connection of these ports, even though the downstream analysis will naturally fail. The data-type agnostic nature of Taverna provides great flexibility, which fosters the inclusion of various Web resources in a workflow, at the cost of guaranteed service compatibility. It is also possible to use local compute grid services as processors in a workflow, an attractive option when the analysis tasks are computationally intensive.

10.4.2 The Results Interface

Clicking the green arrow (the fifth icon from the left in the icon toolbar, under the main menu bar in Figure 10.5) will cause Taverna to validate the workflow. If the validation is successful, the display changes over to the Results view, as shown in Figure 10.6. The main components of this view are (a) the workflow as it is executed (progress meters); (b) the workflow outputs; (c) the list of outputs; and (d) the list of workflow executions.

Before the workflow is executed, any required input parameters are prompted so that the user can provide them. Depending on the workflow, this required input may be a single text value, a specially formatted input file with multiple records, or a spreadsheet. Taverna will automatically iterate

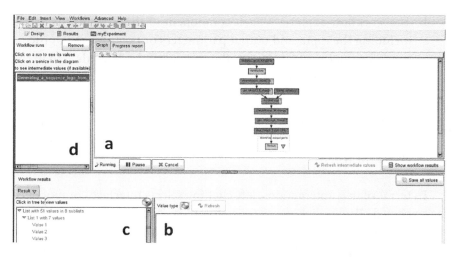

FIGURE 10.6   Taverna interface in Results mode. Main components are: (a) live display of workflow execution progress; (b) display for individual services or total workflow inputs/outputs; (c) input/output data tree for a service or the workflow; and (d) workflow executions instance selector.

through each workflow step from the list of provided inputs, achieving great efficiency compared to manual analysis of genome-scale data.

Workflows do not always have to have explicit inputs from the user. An example would be a workflow that starts with a fixed BioMart query against the *Bos taurus* (cattle) gene database. This fixed query could yield a list of gene names, and these names would be used as input to the rest of the workflow (e.g., a literature search). Running this workflow would require no input from the user. In order to change the behavior of the workflow (e.g., use a different species), the user must edit the settings of the BioMart processor while in Design mode.

While a workflow is executing, for each processor in the workflow display (Figure 10.6a) a progress indicator using color bars is shown. It is important for the efficiency of the analysis of genome-scale data that Taverna automatically forwards each data instance to the next workflow step as it becomes available; the system does not wait for all inputs to complete one task before starting the next analysis step. In Figure 10.6 this is clear, because according to the progress indicators, multiple steps are in progress simultaneously. During and after the workflow execution, intermediate workflow results can be seen by clicking on the desired processor, and then navigating the resulting data tree (Figure 10.6c). Navigation of the data tree requires keeping in mind the nested list processing model for data processing in Taverna. The actual results of interest will not show up in Figure 10.6d until the user has drilled down to the lowest level of the data tree.

Taverna stores workflow execution results in an internal database. It is important to note that in order to ensure that any results of interest are preserved, they have to be saved to a disk before exiting Taverna or they are lost.

10.4.3 Workflow Repository

In general, reusing and modifying an existing workflow as a basis for further genome annotation is a less daunting task than developing a new workflow from scratch. The myExperiment.org Web site (Goble and De Roure, 2007) is a repository for bioinformatics-related workflows, including those written for Taverna and Galaxy. The repository can be searched by keywords, such as the service provider that needs to be accessed for a particular analysis step, or the name of an analysis tool, or the free text descriptions of the workflows, which are provided by the authors. Users may wish to see if an already existing workflow may do the task they seek to perform, to avoid having to do any *de novo* workflow programming. Alternatively, a workflow that performs a related task may exist that would only require some minor modifications using the Galaxy or Taverna editor to achieve the desired results. In either case, the repository includes licensing and attribution information for each available workflow.

10.5 SEAHAWK

Seahawk combines the usability of Galaxy's walled-garden portal approach with the flexibility and power of Taverna's open Web service approach. This combination is achieved in Seahawk by using a Web Service technology called Moby (BioMoby Consortium et al., 2008). Moby consists of an open ontology of domain-specific data types, as shown in Figure 10.7, and a format (XML) to communicate these domain-specific inputs and outputs between Web services. These specialized Web services are called Moby Web services (BioMoby Consortium et al., 2008).

10.5.1 Demonstration-to-Workflow

The basic Seahawk interface (Gordon and Sensen, 2007) is implemented as a Java application, which displays a hypertext rendition of Moby XML data. Users can click on hyperlinks in the document, with each hyperlink corresponding to an XML data instance. The click spawns a menu

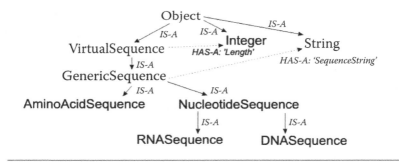

```
<DNASequence articleName='' namespace= 'NCBI_gi' id='154140610'>
  <Integer articleName='Length'  namespace='' id=''>822</Integer>
  <String articleName=' SequenceString' namespace='' id=''>TAAA...AAG</String>
</DNASequence>
```

FIGURE 10.7   Moby ontology. (Top) An example of object inheritance (IS-A) and encapsulation (HAS-A) in the Moby data type ontology. (Bottom) The serialized XML form of an object instance.

of Moby Web services, which can take that particular XML data type as an input. Clicking a service either executes the remote service, or asks for the input of required secondary parameters. Overall, the demonstration experience is very similar to browsing Web pages. Figure 10.8 shows an example. The exception is that hyperlinks can lead to one of several destination documents, that is, service responses. Like in many extant Web browsers, Shift+Click launches new documents into new tabs. This allows users to pursue multiple avenues of investigation simultaneously. Like Galaxy, Seahawk has a built-in workflow-by-example functionality. A Taverna workflow is generated by backtracking from the currently displayed service result (or results if multiple tabs are open). Provenance information, which is embedded in the result documents by Seahawk, is used to backtrack through service calls all the way back to initial user input. The mapping of the demonstration-to-workflow elements is by no means trivial. It takes into account not just the services called but also the data context of each call. This context includes the data filters that were applied and the set of peers for the data used in the demonstration. In the example shown in Figure 10.9, the browsing of four documents during the demonstration stage yields a Taverna workflow with 29 elements: 11 processors, 15 data links, 1 input port, and 2 output ports. Figure 10.9 illustrates Seahawk's preview of this workflow. Various non-Web service processors in the workflow are discussed next.

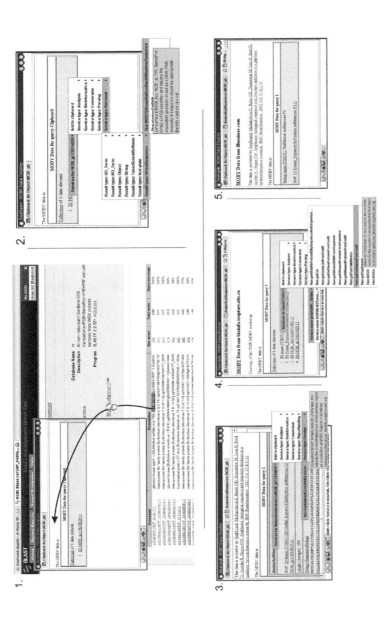

FIGURE 10.8   Seahawk service browsing: From BLAST Web page to sequence and species name tabs in Seahawk. The steps are (1) drag a DB identifier from a Web browser onto the Seahawk clipboard to import the ID; (2) click the ID link and select a service to retrieve the record; (3) Shift+Click the AminoAcidSequence result and select a cross-reference utility (result opens in new tab due to Shift+Click); (4) click the taxon ID to get the species name; and (5) view the result. **(See color insert.)**

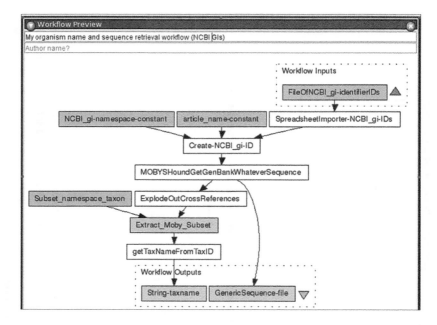

FIGURE 10.9   Seahawk's workflow preview for the Figure 10.8 analysis demonstration. **(See color insert.)**

10.5.2 The Search Widget

Often users will want to retain only a part of the output of the analysis for further processing, especially when they filter large data sets down to subsets, which are likely to be relevant to their particular research question. In Seahawk, a conditional service execution is implemented via a novel search/filter widget. Visually, the behavior is a hybrid of a highlighting search and a gray-out filter. Filtering is essential, so that users will not use data that do not meet the search criteria; these data will simply not be available downstream in the equivalent Taverna workflow. To avoid an inconsistent state, filtering is updated as each letter is typed into the search box. The following three types of conditionals are currently supported in Seahawk:

```
if (cond(x)){f(x)} else {g(x)}
if (cond(x.member1)){f(x.member2)} else {g(x.member2)}
if (cond(f(x))){g(x)}
```

The search condition can be in the format of either a plain string or a regular expression (a powerful pattern matching syntax with wildcards).

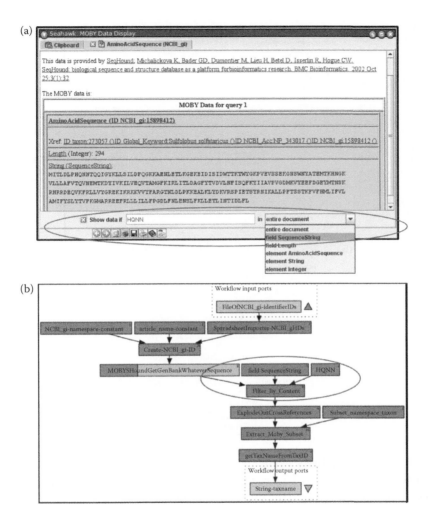

FIGURE 10.10 Seahawk's conditional service execution via a search/filter widget. (a) Filter conditions in Seahawk GUI and (b) corresponding filtering processors in a running Taverna workflow. **(See color insert.)**

Filters are tied to specific documents in Seahawk. Different conditions can therefore be given at each browsing step. "else" conditions are simply selection inversions, based on toggling the keyword "if" to "unless" in the filter widget (Figure 10.10a, bottom left). This fulfills conditionals of type 1 listed above. In terms of the equivalent workflow, a filter processor with two output ports would fork the workflow (Figure 10.10b). In the demonstration, forking is accomplished by calling a service in a new tab (Shift+Click).

FIGURE 10.11 Seahawk's "previous input" service workflow. (a) The "previous input" service selection item available when a data filter is used and (b) workflow components equivalent to the use of this option.

Conditional type 2 filtering is based on the value of a particular member ($x.member1$) of the data object. This is accomplished by selecting from a document-specific drop-down list at the end of the search phrase (Figure 10.10a, bottom right). Any data member ($x.member2$) in a document can be selected for further processing, using the hyperlinks in the data display, unless they are grayed-out by the filter and are therefore inaccessible.

The demonstration of conditional type 3 requires that the user references data item x from the output of $f(x)$. In Seahawk, this is implemented by providing a "for previous input" service menu item when a filter is applied to a service result (i.e., $f(x)$). In Figure 10.11a, selecting the menu item "MOBYSHoundGetGenbankWhateverSequence" is the logical equivalent of the pseudocode:

```
If (getSHoundNeighboursFromGi(15922258) contains "1589"){
    MOBYSHoundGetGenbankWhateverSequence(15922258)
}
```

If no filter is active, the "previous input" option is not displayed, because this would translate into $g(f(x))$, that is, the default interpretation of the service browsing history. The set of Taverna workflow activities required to implement condition type 3 is nontrivial (Figure 10.11b). These processors include Java code and it would almost certainly be beyond the capabilities of a novice user to create the equivalent of these via programming.

Other somewhat unusual aspects of the search/filter behavior in Seahawk are page-specificity and job-level filtering. Page specificity implies that the filter widget is tied to the given document. This means that when a user navigates away from a page, the widget disappears. Upon reentry to the given page, it will reappear. Hiding the search/filter widget disables the filtering, whereas showing the widget again enables the last known filter criteria. This stateful, page-specific behavior is essential for different conditions to be applied at different stages of the workflow. Statefulness also means that the output from one service can be filtered multiple ways. This option for forking is a variation on the if–else concept. It is accomplished by launching subsequent services in new tabs after each search criterion has been applied.

Job-level filtering reflects the for-each functionality in Seahawk. In a document containing results for 10 inputs (i.e., 10 "jobs"), the filtering decision is to either allow or disallow the subsequent processing of each job individually; they are independent results. Having lists of jobs helps to guide the analysis demonstration, because users receive immediate feedback on the correctness of positive and negative examples. This is in contrast to traditional programming strategies, where the user must surmise the correctness of the filter expression for arbitrary data.

10.5.3 Data Filters and Labels

Data filter processors found in Seahawk-generated workflows serve to ensure data type safety, which Taverna normally would not. For example, in Figure 10.9 the "Extract_Moby_Subset" processor ensures that only the data contained in the taxon namespace are passed from

"ExplodeOutCrossReferences" to the "GetTaxNameFromTaxId" service. Without this filter, "GetTaxNameFromTaxId" could also receive an NCBI_gi ID as input (another Xref in Figure 10.8). As the service cannot interpret an NCBI_gi at all, Seahawk makes a concerted effort to ensure workflow data type safety, in order to minimize problems for the user when executing the workflow in Taverna.

The data filters are implemented as BeanShell scripts. These scripts are essentially small Java programs. The scripts can be inspected by the user in the Taverna user interface but are relatively opaque to novice programmers. A design decision in Seahawk was to have the BeanShell filters (in beige, e.g., "Extract_Moby_Subset") generic and parameterized. In this way, for example, the "taxon" filter could be easily identified and changed without writing any additional Java code.

Building a type-safe workflow is not sufficient; the workflow must also be able to visually represent the story of the analysis. This can help users understand the correspondence between their demonstration actions and the resulting components of the workflow. This may in turn help users learn the use of Taverna. Seahawk leverages the semantic information provided by Moby Web services; these domain-centric labels are attached to the various processors of the workflow. In Figure 10.9 each output port is named according to its data type (e.g., String) and corresponding Web service output field name (e.g., taxname). This is in contrast to the Galaxy examples, where data may be simply labeled "Output from Step 2."

Seahawk also takes advantage of Taverna's T2Flow language support for several types of annotations. Titles and free text descriptions can be added to a workflow when being exported from Seahawk (Figure 10.9). Semiformal annotations are also provided; in case a user does not know what to use as input to the workflow, Seahawk provides an example value. The example annotation uses the input value from the demonstration stage.

10.5.4 Taverna Enactment of Seahawk-Generated Workflows

A unique feature of Seahawk among programming-by-demonstration systems is that it generates a preview of the workflow before saving it. The workflow preview image is currently generated by an automated call to a remote server. Without downloading and running Taverna, the preview allows users to quickly go back to adjust their Seahawk browsing or filtering if the workflow does not capture their intent.

For example, a user examining Figure 10.9 would quickly realize his or her mistake if all of the analysis was done in one tab. As a rule, there are as many output ports in a workflow as there are open Seahawk tabs. In the example, the workflow preview would contain one output instead of two. To fix this, the user could go back and launch the cross-reference search in a new tab instead. This would achieve a workflow with both expected outputs. Once the user is satisfied with the workflow, it can be saved and subsequently loaded into Taverna.

Large-scale genomic data analysis is often somewhat trial-and-error; therefore making it easy to modify parameters in Taverna would be very useful to the novice user. These parameters, for example, might affect behaviors such as data filtering and decomposition. To this end, behavior parameters are clearly marked as external constants in Seahawk workflows. Minimizing processor redundancy can help alleviate workflow crowding. This involves Seahawk tracking: (1) when an input has been used more than once; (2) when a filter is being applied multiple times; (3) when the same object member is extracted for further processing multiple times; and (4) for filter negations (if–else).

Since the data and services are semantically typed, Seahawk can generate meaningful labels for otherwise anonymous nodes in a workflow, such as inputs and outputs. Example values gleaned from the demonstration are also added as metadata to the workflow. By providing fields for additional metadata in the workflow preview (e.g., Figure 10.9), Seahawk promotes readability and reusability of the generated workflows.

The actual enactment of the workflow is largely but not entirely outside the control of Seahawk, because the workflow is run eventually using Taverna. Since Taverna is largely data-format agnostic, a user's inability to input data correctly to the generated workflow can be a serious barrier to success. To address this barrier, Seahawk includes extra processors in the workflows, simplifying data input. For example, a spreadsheet importer (visible at the top of Figure 10.12) allows users to specify their input lists within the familiar environment of rows and columns in Excel.

Hints are provided in the Taverna "Port description" to tell the user what data to put in which columns (Figure 10.12). Without this importer, the workflow would require Moby XML as input, something users are unlikely to have on hand for their data of interest. A spreadsheet may contain a list of GI identifiers, to which the workflow will be applied. For example, this input list may be the result of a BioMart query on the Ensembl genome annotation page for the human genome. The Taverna workflow would allow the user to augment the Ensembl annotation for a large number of genes, without manually running each GI through a series of Web forms.

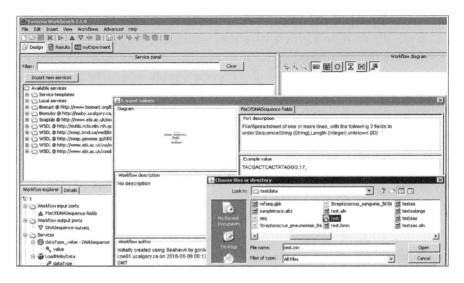

FIGURE 10.12 Input dialog for a Taverna workflow generated by Seahawk. A spreadsheet with the format described in the "Port description" box is expected.

The ability to import spreadsheets also provides an opportunity for users to apply whatever spreadsheet programming knowledge they have. This can be used to supplement the analysis done in the demonstration or workflow. This spreadsheet plus demonstration hybrid approach to analysis automation provides an avenue for end-users to learn automation of genomic analyses in a practical way.

10.6 CONCLUSION

Genome annotation tools tend to produce a fairly fixed set of analysis outputs. Depending on the research question at hand, the available data may be insufficient to properly answer the research question. The custom analysis of genomic scale data by nonprogrammers is becoming viable through the evolution of workflow and workflow-by-example tools. Web-based workflow tools can usually be used by anyone without requiring a large local informatics infrastructure, though local installation options are available for some tools. This can increase the efficiency (Taverna) and allow for the inclusion of additional tools (Galaxy) that might not be available in the publicly accessible versions of the tools. Workflows provide several advantages over manual analyses: efficiency, research communication, process documentation, results reproducibility, keeping results up to date, and the potential reuse and modification of components for related tasks.

REFERENCES

Bhagat, J., Tanoh, F., Nzuobontane, E., et al. 2010. BioCatalogue: A universal catalogue of web services for the life sciences. *Nucleic Acids Res.* 38:W689–W694.

BioMoby Consortium, Wilkinson, M.D., Senger, M. 2008. Interoperability with Moby 1.0—It's better than sharing your toothbrush! *Brief. Bioinform.* 9(3):220–231.

Chagoyen, M., Kurul, M.E., De-Alarcon, P.A., Carazo, J.M., Gupta, A. 2004. Designing and executing scientific workflows with a programmable integrator. *Bioinformatics* 20(13):2092–2100.

de Knikker, R., Guo, Y., Li, J., et al. 2004. A Web services choreography scenario for interoperating bioinformatics applications. *BMC Bioinformatics* 5:25.

Fielding, R.T., Taylor, R.N. 2002. Principled design of the modern Web architecture. *ACM Trans. Internet Technol.* 2(2):115–150.

Flicek, P., Aken, B.L., Ballester, B., et al. 2010. Ensembl's 10th year. *Nucleic Acids Res.* 38(Database issue):D557–D562.

Goble, C.A., De Roure, D.C. 2007. myExperiment: Social networking for workflow-using e-scientists. In *Second Workshop on Workflows in Support of Large-Scale Science*, eds. E. Deelman, I. Taylor, 1–2. New York: ACM Press.

Goecks, J., Nekrutenko, A., Taylor, J.; Galaxy Team. 2010. Galaxy: A comprehensive approach for supporting accessible, reproducible, and transparent computational research in the life sciences. *Genome Biol.* 11:R86.

Gordon, P.M.K., Sensen, C.W. 2007. Seahawk: Moving beyond HTML in Web-based bioinformatics analysis. *BMC Bioinformatics* 8:208.

Hoon, S., Ratnapu, K.K., Chia, J.-M. 2003. Biopipe: A flexible framework for protocol-based bioinformatics analysis. *Genome Res.* 13(8):1904–1915.

Lee, S., Wang, T.D., Hashmi, N., Cummings, M.P. 2007. Bio-STEER: A semantic Web workflow tool for grid computing in the life sciences. *Future Gener. Comput. Syst.* 23(3):497–509.

Ludaescher, B., Altintas, I., Berkley, C., et al. 2006. Scientific workflow management and the Kepler system. *Concurr. Comput.: Prac. Exp.* 18:1039–1065.

Oinn, T., Addis, M., Ferris, J., et al. 2004. Taverna: A tool for the composition and enactment of bioinformatics workflows. *Bioinformatics* 20 (17):3045–3054.

Pautasso, C., Alonso, G. 2005. The JOpera visual composition language. *J. Vis. Lang. Comput.* 16(1–2):119–152.

Rice, P., Longden, I., Bleasby, A. 2000. EMBOSS: The European Molecular Biology Open Software Suite. *Trends Genet.* 16(6):276–277.

Shah, S., He, D., Sawkins, J., et al. 2004. Pegasys: Software for executing and integrating analyses of biological sequences. *BMC Bioinformatics* 5(1):40.

Stajich, J.E., Block, D., Boulez, K., et al. 2002. The BioPerl toolkit: Perl modules for the life sciences. *Genome Res.* 12:1611–1618.

Tang, F., Chua, C.L., Ho, L.-Y., Ping Lim, Y., Issac, P., Krishnan, A. 2005. Wildfire: Distributed, grid-enabled workflow construction and execution. *BMC Bioinformatics* 6:69.

CHAPTER 11

Analysis Pipelines for Next-Generation Sequencing Data

11.1 INTRODUCTION

The recent advance in high-throughput sequencing technologies and the availability of next-generation sequencing (NGS) machines have resulted in the creation of massive amounts of sequencing data in a relatively short period of time. As of early 2012, the three most widely used sequencing platforms were Roche 454 Life Sciences Genome Sequencer (www.454. com), the Illumina Solexa Genome Analyzer (GA) (www.illumina.com), and the Applied Biosystems (ABI) SOLiD system (www.appliedbio-systems.com). There are also newer platforms, such as Polonator G.007 (www.polonator.org), Helicos Biosciences Heliscope (www.helicosbio. com) and the Pacific Biosciences RS (www.pacificbiosciences.com), that are gaining popularity as yet another group of massively parallel sequencing machines. The sequencing methods of these sequencers in terms of their biochemistry, cost per base, processing speed, strengths, and weaknesses have been recently reviewed and compared (Mardis, 2008; Metzker 2010; Pettersson et al., 2009; Shendure and Ji, 2008).

There are several features of these new sequencing technologies that have significant implications in data analysis, starting with genome reconstruction. Most importantly, short reads are produced in the range of around 400 base pairs from 454 sequencers, and 100 base pairs or less from Solexa and SOLiD machines. These short reads provide little information per read, causing difficulty in assembly, especially in handling those reads coming from repeat regions within the genome. To compensate for this shortfall, the assembly of NGS sequence reads requires a high degree of coverage over the size of the genome to be assembled, which results in a tremendous increase in read numbers that are produced for any particular genome. A single machine run can now yield 500 million individual sequence reads!

Some of the new technologies are known to produce some characteristic systematic errors, such as difficulty estimating the length of homopolymer regions (i.e., single base repeats) and increasing base call errors toward the 3′ ends of reads. All of the aforementioned sequencing technologies include an estimate of the confidence in each base call as a quality value. These characteristics of NGS data, which can be summarized as escalated ambiguity during assembly and exponential increase in the sheer amount of information to be processed downstream, pose computational challenges for NGS data analysis pipelines. More computationally intensive algorithms to cope with the ambiguity are needed, together with more computing power and ample storage space to accommodate sequence and other types of genomic data. The computational challenges of NGS data analysis have been recently reviewed (Li et al., 2011; Pop, 2009).

11.2 GENOME SEQUENCE RECONSTRUCTION

The first important task in most NGS data analysis workflows is the reconstruction of as complete a genomic sequence as possible, beginning from the short individual reads, which by themselves are hardly meaningful. When it is known which organism the sequencing reads came from, they can be aligned to a previously sequenced and annotated reference genome, if one of the same or very closely related species is available. The goal of the alignment process is to determine for each read the most likely source location within the reference genome sequence from which the read could have originated. The alignment approach is only appropriate for resequencing or comparing genetic profiles of closely related organisms within specific species. For example, a study on genetic variation between *Arabidopsis thaliana* strains has been performed by NGS read alignment (Ossowski et al., 2008).

When the organism being sequenced is not similar to any existing sequenced genomes, its genome sequence must be reconstructed from the collection of reads *de novo*. Although not computationally intractable, *de novo* genome assembly is much more demanding than alignment, with regard to algorithm complexity and computing resources. Because of this difficulty, *de novo* assembly of short reads has mostly been applied to the reconstruction of bacterial genomes (Butler et al., 2008; Chaisson and Pevzner, 2008; Warren et al., 2007) or mammalian bacterial artificial chromosomes (Zerbino and Birney, 2008). It should also be noted that alignment and assembly approaches are not always used in a mutually exclusive way. If certain regions of the new genome sequence do not align to anywhere in the reference genome, those regions need to be assembled *de novo*.

11.2.1 Alignment to the Reference Genome

Although earlier alignment tools such as BLAST (Altschul et al., 1990), Clustal (Chenna et al., 2003), and T-Coffee (Notredame et al., 2000) have been popular for many years and are still in widespread use, their primary purpose is either to search for relatively small homologous sequences in large sequence databases or to align a given set of sequences for the subsequent inference of their evolutionary relationships. In contrast, short read alignment for NGS data requires mapping of those reads to the reference genomic sequence of the same or very closely related species, with the goal of detecting structural variation or guiding genome reconstruction. Therefore, unlike classic alignment algorithms, most short-read alignment algorithms assume that nonmatching bases stem from variation within a particular species rather than evolutionary substitutions. This subtle but important difference in underlying assumptions allows for implementations of relatively fast alignment algorithms for NGS reads, since fewer mismatches are normally permitted and initial candidate match locations in the reference sequence can be quickly narrowed to a manageable number.

Most short-read alignment algorithms seek to strike a balance between processing speed and alignment accuracy. As a result, these algorithms generally adopt a two-step strategy, where the first step is devoted to rapidly finding a reasonably small number of potential locations in the reference genome to map each read by employing simple heuristics, such as exact match in a preset number of bases in the beginning of a read. Once the candidate mapping locations are identified, a more accurate but slow alignment algorithm, such as a variant of the Smith–Waterman algorithm (Smith and Waterman, 1981), is applied on the set of sequences in a second step. The stringency of the heuristics applied in the first step can normally be controlled by the users via a set of parameters, such as the number of the bases to match and the number of mismatches allowed in those bases.

Recently, there has been proliferation of new software tools for NGS short-read alignment. This wave of development was mainly motivated by the inadequacy of traditional alignment algorithms to efficiently handle the massive amounts of short reads, as well as increasing genome sizes for which sequencing projects are undertaken. The previous generation of tools, for the most part, had been designed to deal with smaller numbers of longer reads. Some of the most widely used NGS alignment tools include Bowtie (Langmead et al., 2009), SOAP (Li et al., 2009), BWA (Li and Durbin, 2009), SHRiMP (Rumble et al., 2009), mrFAST (Alkan et al., 2009), mrsFAST (Hach et al., 2010), ZOOM (Lin et al., 2008), SSAHA2 (www.sanger.ac.uk/resources/software/ssaha2) (Ning et al., 2001), and Mosaik (http://bioinformatics.bc.edu/marthlab/Mosaik). The performance of several of these algorithms was recently compared, with respect to accuracy based on numbers of correctly and incorrectly mapped reads and unmapped reads, as well as runtime (Ruffalo et al., 2011).

11.2.2 *De Novo* Assembly

One of the biggest challenges in reconstructing a genome sequence from short reads without a reference genome is the need to resolve repeats, especially when the reads are shorter than the overall repeat length or paired-end reads are not available. Increasing the read coverage by oversampling has been the main strategy to cope with this difficulty, which has resulted in very large numbers of reads that assembly algorithms have to deal with in a single assembly. The degrees of coverage over different genomic regions are not always uniform, as systematic variability specific to local genomic regions exists, depending on the sequencing technology used. For example, a more than 100-fold difference in per-base coverage among Roche 454, Illumina GA, and the ABI SOLiD technologies for the same human genomic regions was reported (Harismendy et al., 2009). Because these variable coverage patterns are specific to the respective sequencing technologies, studies on combining read data from multiple sequencing technologies have been performed and reported to yield improved *de novo* assembly results in comparison to those obtained using only the data from a single sequencing technology (Aury et al., 2008; Reinhardt et al., 2009). As a general rule, it is always advisable to combine long- and short-read NGS technologies for *de novo* sequencing projects.

NGS assembly algorithms use a computational strategy to cope with the overwhelming complexity of processing large numbers of reads, especially in the usual first step of calculating overlaps among the reads. In the case of fully exhaustive overlap comparison, each read needs to be compared with all other reads, resulting in combinatorial explosion. For efficiency, most *de novo* assembly algorithms utilize the notion of *k*-mer to a certain extent. Instead of finding overlaps between reads, a set of reads sharing a same *k*-mer is found and only those reads that share the same *k*-mer are compared against one another to identify overlaps. This strategy drastically reduces the computational requirement when compared to all-to-all overlap calculation of reads. However, the value of *k* should be carefully chosen, such that true overlaps are not missed and false-positive overlaps are minimized. In addition, most assembly algorithms also use compact mathematical graphs as underlying representation of reads, contigs, and their relationships.

A number of *de novo* assembly software tools have been released during the last few years. The first group of assembly tools used are called "greedy algorithms," where a given read is extended by adding the next highest-scoring overlapping read, and this basic operation is applied to joining contigs as well. Tools in this category include SSAKE (Warren et al., 2007), SHARCGS (Dohm et al., 2007), and VCAKE (Jeck et al., 2007). The next major assembly strategy is overlap–layout–consensus (OLC), which works in three steps: (1) overlaps among all reads are calculated, (2) which are then laid out on a graph, where overlapping read nodes are connected, and (3) finally the consensus sequence is generated by analyzing the graph. For example, Newbler (Margulies et al., 2005), Celera Assembler (Myers et al., 2000) and its extension for 454 data called CABOG (Miller et al., 2008), ARACHNE (Batzoglou et al., 2002), and Edena (Hernandez et al., 2008) are assembly tools based on the OLC scheme. Yet another strategy uses de Bruijn graphs (DBGs), where reads are first cut into *k*-mers, a DBG is formed using all the *k*-mers, and then the genome sequence is inferred from the DBG. Assembly software such as EULER-USR (Chaisson et al., 2009), Velvet (Zerbino and Birney, 2008), ABySS (Simpson et al., 2009), ALLPATHS-LG (Gnerrea et al., 2011), and SOAPdenovo (Li et al., 2010) belong to this last category. Several reviews of assembly algorithms and tools have been published (Flicek and Birney, 2009; Li et al., 2011; Miller et al., 2010; Shendure and Ji, 2008).

11.3 ANALYSIS PIPELINES: CASE STUDIES

Over the last few years, the Visual Genomics Centre at the University of Calgary has developed several pipelines for NGS data analysis. These pipelines, which are essentially tool integration systems tailored to specific genome projects and their analysis needs, exemplify different genome analysis applications for problems of high socio-economical impact that require NGS data to be solved.

11.3.1 16S rRNA Analysis

A small subunit ribosomal RNA (SSU rRNA) data analysis pipeline was developed to perform microbial community analysis in oil sands and coal beds. This pipeline follows the metagenomics analysis principles outlined in Section 4.3. Figure 11.1 shows the flow of data in the pipeline. Raw 454 reads of 16S rRNA are subjected to stringent systematic checks in order to remove low quality reads and minimize the effect of sequencing errors (Huse et al., 2007). Filtered sequences include ones that (1) do not perfectly match the adaptor and primer sequences, (2) contain ambiguous bases, (3) have an average quality score below 27, (4) contain homopolymers of a length greater than 8, and (5) are shorter than 200 base pairs after clipping off the primers.

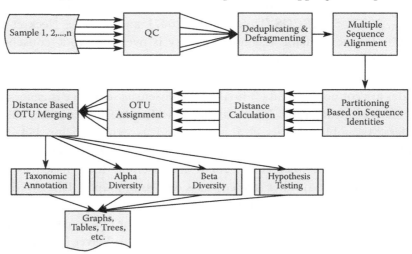

FIGURE 11.1 16S rRNA analysis pipeline developed at the Visual Genomics Centre, University of Calgary. 16S rRNA samples are quality checked, partitions are created from unique sequences, OTUs are generated by clustering, and several diversity indices and taxonomic annotations are generated. (Courtesy of Xiaoli Dong.)

The subsequent microbial community analysis is based mainly on Mothur (Schloss et al., 2009), a widely used 16S rRNA data analysis package. As Mothur in itself is not capable of handling NGS data, two major steps have been developed for the pipeline to handle larger data sets. In the first step, the data set is collated prior to sequence alignment to include only unique sequences. The clustering of redundant overlapping sequences is done by the CD-HIT-EST program (Li and Godzik, 2006), at a 99% identity threshold. Only the representative sequences generated by the CD-HIT-EST program in each cluster are included in the final alignment. In the second step, the remaining unique sequences are preclustered by CD-HIT-EST again using 80% identity threshold to partition the data set into a series of smaller subsets. Then the distance calculation and clustering is performed in parallel over the smaller data sets instead of over the entire data set, because this step forms the major processing bottleneck in Mothur.

The pipeline then uses Mothur to generate operational taxonomic units (OTUs) at 3% and 5% distances using the average linkage algorithm (Schloss and Westcott, 2011). After grouping sequences into OTUs, several alpha diversity indices (Whittaker, 1972) as well as the total numbers of OTUs are calculated for each sample. To explore potential relationships between microbial communities from different environments, sample dissimilarities are calculated and a tree is built, which can subsequently be visualized. Differences and similarities among samples are also explored by an ordination analysis, which reduces the dimensionality of the data to be compared from the number of OTUs generated to a small number that can be easily visualized. Nonmetric multidimensional scaling (Prentice, 1977) and principal component analysis (Wold et al., 1987) are the two methods that are being used in the pipeline for this purpose.

Finally, the OTUs are given taxonomic annotation by using three methods. A BLAST search against the nonredundant SSU reference data set of SILVA102 (Pruesse et al., 2007) using the Tera-BLAST algorithm on a TimeLogic DeCypher system (Active Motif, Inc.) is performed to identify similar sequences. The sequences are also searched against the RDP database and the SILVA reference database (www.mothur.org/wiki/Taxonomy_outline), respectively, using the RDP classifier, which is implemented as part of Mothur. The results of these taxonomic classification methods are provided as plain-text files and also as MEGAN-compatible files for visualization using MEGAN (Huson et al., 2007).

11.3.2 Targeted EST Assembly

In order to analyze the genetic material of plants to determine specific gene sequences and select candidate or targeted genes that play a key role in their metabolic pathways, a targeted expressed sequence tag (EST) assembly pipeline has been developed. ESTs are reads of sequences obtained from sequencing cDNA libraries (Nagaraj et al., 2006). As cDNA is complementary to mRNA, ESTs characterize the expressed genes in specific tissues at certain developmental stages. ESTs have become an essential tool to study gene expressions in tissues with regard to different developmental stages and a variety of environmental factors. They have been used in large-scale transcriptomic studies to aid gene discovery, enhance genome annotation, determine the structure of genes, and discover splicing patterns (Wang et al., 2009).

The general transciptomic analysis pipeline when using EST sequences consists of mRNA preparation, cDNA library construction, read retrieval from sequencer, read quality trimming and filtering, read clustering, assembly, and annotation of the contigs generated by the assembly program. Because of the large quantities of the reads, it is still a big challenge to assemble the reads to produce a complete representation of all the expressed genes. In order to increase the computing efficiency in EST analysis, target-restricted assembly methods have been used to reconstruct sequence ortholog regions in low-coverage sequences (Bainbridge et al., 2007) and a single lane of Illumina sequences for genes of relevance to phylogeny reconstruction (Johnson et al., 2010). In target-restricted assembly pipelines, the target protein-coding sequences of interest are first identified and built into a database. Then BLAST-based algorithms are normally used to retrieve all of the reads closely related to the protein-coding targets of interest. Those reads are then assembled to generate complete full-length coding sequences of the genes of interest.

In addition to BLAST-based search tools, searching algorithms based on hidden Markov models (HMMs) also have been widely used in sequence alignment and gene discovery. HMMER (http://hmmer.janelia.org) (Finn et al., 2011), an HMM-based sequence homology search tool, has been used for searching homologs of protein sequences and aligning protein sequences. Compared to BLAST, HMMER can detect more remote homologs more accurately (Finn et al., 2011). By combining BLAST and HMMER, a computationally inexpensive assembly pipeline was developed to recognize and assemble targeted

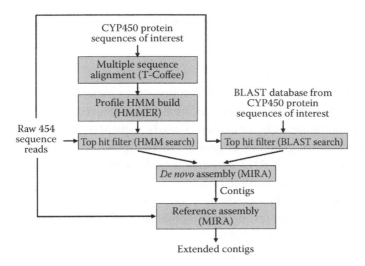

FIGURE 11.2   Targeted EST assembly pipeline developed at the Visual Genomics Centre, University of Calgary. Protein sequences of interest are used to build HMM and BLAST databases, the raw reads are used as queries, the hits are assembled into contigs to serve as the reference assembly and the raw reads are assembled again with the reference. (Courtesy of Mei Xiao and Ye Zhang.)

protein-coding regions of interest from cDNA sequences generated by NGS technologies.

Figure 11.2 shows the targeted assembly pipeline, where the targets are cytochrome P450 protein sequences. In order to use HMMER to search for the related 454 reads, we first used T-Coffee (Notredame et al., 2000) to do a multiple alignment on all the P450 protein sequences. The aligned protein sequences were fed to HMMER to build a profile HMM-based database. The original 454 reads were used as queries to search the profile HMM database. In addition to the HMM-based search, we also took all the P450 protein sequences and built a BLAST database. The original 454 reads were again used as queries to search the BLAST database using the *tblastx* program to search through our protein database using a translated nucleotide query. The results from both the profile HMM-based search and the BLAST search were combined. A filter was applied to the combined 454 read sequences to remove any redundant sequences. The remaining nonredundant 454 read sequences were used as input to the MIRA assembly program (Chevreux et al., 2004). MIRA was used to assemble the 454 reads related to the targeted protein sequences into longer contigs. Those contigs were again used as a reference and MIRA was used a second time to assemble the original 454 reads with a reference data set.

11.3.3 Gene Prediction

To search for better enzymes to decompose plant material into biofuel, there is a need to predict the accurate locations, exon structure, and functions of protein-coding genes in a number of newly sequenced as well as previously published fungal genomes. A gene prediction pipeline integrating *ab initio* prediction tools with RNA-Seq and homology evidence was developed to predict gene models in fungal genomes. The design of the pipeline allows results from individual prediction tools or evidence sources to be easily added or removed. By testing several possible combinations, a minimal set of components was selected that works well together and produces high-quality gene models within a week, given a genomic sequence and RNA-Seq reads.

Two *ab initio* prediction tools with demonstrated good results for fungal genomes, GeneMark ES (Ter-Hovhannisyan et al., 2008) and AUGUSTUS (Stanke et al., 2008), are used as the main tools in the gene prediction pipeline. These two tools are mutually complementary in that GeneMark ES does not need a training set of genes from the target genome, whereas AUGUSTUS requires training models for genes from the target genome or a closely related genome. GeneMark ES uses an iterative training strategy on the genomic sequence itself to predict a set of gene models. The AUGUSTUS training set can be used to optimize the transition and emission probabilities of the HMM as well as several other parameters of the prediction tool. Experimental hints about gene locations can also be used in AUGUSTUS to improve its prediction accuracy, such as those from protein, EST, or RNA-Seq alignments. In fact, RNA-Seq information is used at several steps of the pipeline. It is used for the generation of the initial models, which are used for the training of the HMM in AUGUSTUS, as well as for providing additional hints to AUGUSTUS while creating a set of predicted gene models. The final step of the pipeline, where the output models are scored and the best ones are selected, also uses this information.

Figure 11.3 shows the data flow in the pipeline. The pipeline uses full-length transcripts, which were assembled into contigs from RNA-Seq reads using Velvet (Zerbino and Birney, 2008) to generate a set of initial gene models (contig training set) for training the HMM-based gene predictor AUGUSTUS (Stanke and Waack, 2003). AUGUSTUS is trained using the alignment of RNA-Seq contigs to the genome by exonerate (Slater and Birney, 2005). A consensus set of predicted gene models is generated from

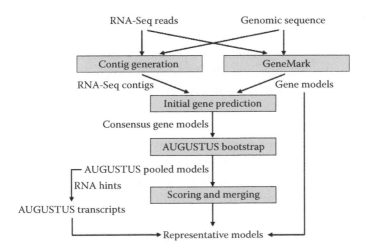

FIGURE 11.3 Gene prediction pipeline developed at the Visual Genomics Centre, University of Calgary. GeneMark-generated models and RNA-Seq read contigs are used to create prediction sets, which are used to train the *ab initio* gene predictor AUGUSTUS. The AUGUSTUS predictions are scored and selected to produce the final set of gene models. (Courtesy of Mostafa Abdellateef.)

AUGUSTUS using the contig training set and the self-training predictor GeneMark. This set is used to train AUGUSTUS yet again, in preparation for the second stage of predictions. Hints based on RNA-Seq data are also used in this stage to improve the quality of AUGUSTUS predictions. Gene models produced in this stage are scored for agreement with intron locations, which were inferred from the RNA-Seq data. This scoring stages tests for translatability into predicted proteins and for homology of the predicted proteins to known proteins. The set of nonoverlapping models with the highest score constitutes the final result of the pipeline.

11.4 NEXT-GENERATION GENOME BROWSING

11.4.1 Integration of Different Types of Genomic Data

Most genome browsers described up to this point can only display genomic sequences and their various annotations from multiple sources. Although these browsers can visualize multiple tracks of information, there is an ever-increasing need to visualize different types of data simultaneously to gain intuitive understanding. For example, with the generation of large data sets from microarray experiments, there is an increasing need to generate views of both genomic and expression data within a common visual context and to perform a visually guided interpretation of the data. Traditional gene expression visualization tools normally display expression values as a heat map using color gradients to represent the strengths or weaknesses of expression. For example, the TIGR MultiExperiment Viewer (MeV) (Saeed et al., 2003) is a publicly available gene expression analysis and visualization tool. With this type of visualization, however, it is difficult for the researcher to see and interpret the gene expression profile within a genomic context, as each gene is represented simply as a text symbol.

In a similar vein, cancer genomics entails searching for the genes and their mutations that contribute to the development of a cancer cell in order to analyze the alterations that occur in the genome of a patient's cancer cells. Tumor samples are analyzed using microarray and high-throughput DNA sequencing technologies to generate genomic data with clinical significance, such as insertions, deletions, and the copy numbers of genes. Large clinical trials of this kind produce a huge amount of different types of molecular data that needs to be organized to provide a unified view. There are several genome browsers that were developed for integrating genomics data with clinical data, especially gene expression data and cancer genomics data. These browsers are capable of simultaneously displaying genome-scale clinical measurement values for sets of samples in reference to genomic coordinates.

For example, TIGR MeV has been integrated into Bluejay (Soh et al., 2012; Soh et al., 2008) at the source level in order to extend its functionality to enable the display of gene expression levels within the genomic context. Bluejay internally represents the genomic data, along with gene expression values, as a document object model. The results of the gene cluster analysis performed by TIGR MeV can also be displayed on the whole genome, with each cluster represented in a unique color.

Figure 11.4 illustrates the interplay between the TIGR MeV module and Bluejay for visualizing genome sequence data and gene expression data simultaneously. Microarray data integration in Bluejay adds the capability of showing gene expression values and genes side by side within a whole genome, so that the association between genes and their expression values is visually evident. For a biologist, it becomes much easier to interpret gene expression analysis results than when the results were displayed outside of the genomic context, as in expression images, heat maps, or tables of numeric values.

The University of California, Santa Cruz (UCSC) Cancer Genomics Browser (Zhu et al., 2009) is an extension of the UCSC Genome Browser,

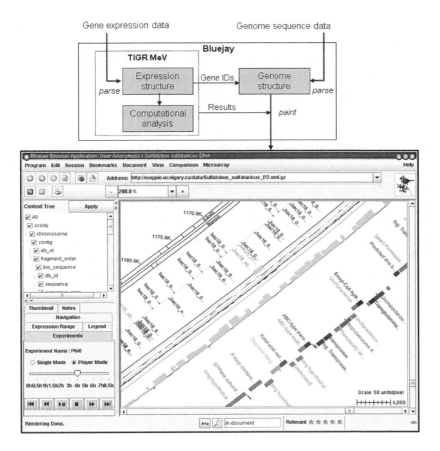

FIGURE 11.4 Integration of gene expression analysis with genomic data in Bluejay. Gene expression data parsing and analysis are done by TIGR MeV. The expression and analysis data are combined with the genome data to produce a unified visual representation, where the genes and their expression/analysis values are displayed side by side. **(See color insert.)**

which displays data and annotations in parallel tracks. The Cancer Genomics Browser represents data as heat maps, in which colors represent the values of key variables, the horizontal axis represents genomic positions, and the vertical axis represents a stack of genomewide measurements from multiple samples. Genomic and clinical data are displayed alongside each other, and tools are provided so that users can sort, filter, and group the clinical data based on features of interest and run statistical analyses. This makes it easy for the user to identify apparent patterns, such as frequent loss of some genes, across different samples. The browser site (http://genome-cancer.ucsc.edu) hosts a large amount of publicly available cancer genomics data.

The Integrative Genomics Viewer (IGV; www.broadinstitute.org/igv) (Robinson et al., 2011) has been developed at the Broad Institute with the goal of helping users to simultaneously visualize and analyze different kinds of genomic data by bringing them together into a single unified view. On top of the genomic sequences, additional data layers can be displayed, including gene expression, sequence alterations or mutations, and copy number variation. Figure 11.5 shows an example of using IGV, where two different mapping results are laid out together vertically. The upper part depicts the mapping of contigs while the lower part shows the mapping of raw reads, enabling the user to view the reference genome, assembled contigs, and raw reads in the same genomic context. A variety of display options are available for users, such as viewing data as a heat map, histogram, scatter plot, or other formats of choice. IGV was designed

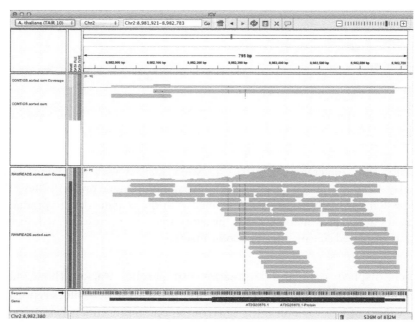

FIGURE 11.5   Integrative Genomics Viewer (IGV) display of contigs and raw reads mapped to the *Arabidopsis thaliana* reference genome is divided into chromosome ideogram, tracks of gene expression sample data, and gene feature track. Typically each horizontal data track represents a single sample or experiment, whose values are compared vertically with other data tracks and also mapped to the feature track.

to be general purpose, so that it is possible to integrate mostly any kind of genomic data, rather than as a specialized tool that can be used to display a specific type of data. Users can also zoom and scroll at any viewing scale, and the display tracks can be reorganized based on selected regions and genomic features. Data can be input in various commonly used file formats depending on the type of data.

11.4.2 Decentralization

Most Web-based genome browsers are based on a centralized server model, in which primarily the server-side processing is used for database access and image rendering. At the request of the user, information flows from the databases to the genome browser, which generates the display as images and transmits them to the user for viewing. As the size of the data set increases and the browser functionality advances, this communication between the server side and the client side can become the bottleneck in the whole genome browsing experience as well as lead to the overloading of the server. There are two main ways to tackle this problem: more client-side processing and use of distributed resources.

The amount of client-side processing can be increased to reduce the computational burden of the server as well as to minimize network traffic. An example for this is increasing use of the Asynchronous JavaScript and XML (AJAX; http://ajax.org) technologies, a collection of interrelated client-side Web development techniques. AJAX allows genome browsers to distribute the workload better by allowing Web applications to retrieve data from the server asynchronously, without disrupting the behavior of the client-side display. This usually results in smoother navigation of a genome. JBrowse (Skinner et al., 2009) is an AJAX-based version of GBrowse. JBrowse uses AJAX to reduce the server load, generate smooth scrolling, and provide intuitive semantic zooming. JBrowse contains all the existing features of GBrowse, while using the client-side Web browser to do most of the image rendering and display. Several other genome browsers, including Anno-J (www.annoj.org), X:map (Yates et al., 2008), and Genome Projector (Arakawa et al., 2009), use similar technologies to improve server response times and allow smooth navigation through genomes.

Biologists have access to numerous distributed resources, such as biological databanks, software applications, and Web services. Nevertheless, when they need to investigate a particular organism or a set of related organisms, these resources will prove useful only if they are integrated and linked together in a way that is conducive to the study. There are established methods to analyze various types of data on their own, but the challenge is to unify these analyses into a coherent model of an organism. For example, Bluejay offers a visually guided data exploration environment that can interoperate with other computational tools as well as integrate heterogeneous types of biological information using its unified visual representations. Seahawk is an application that provides access to Moby-compliant biological Web services (Gordon and Sensen, 2007), which was described in greater detail in Chapter 10. Seahawk is incorporated into Bluejay as a single Java Archive file without requiring the use of specifically designed API for launching it from within Bluejay to access Web Services. When a new release of Seahawk becomes available, the JAR file can simply be replaced. This approach makes effective use of distributed computational resources on demand rather than centrally installing all the services on a genome browser server.

REFERENCES

Alkan, C., Kidd, J.M., Marques-Bonet, T., et al. 2009. Personalized copy number and segmental duplication maps using next-generation sequencing. *Nat. Genet.* 41:1061–1067.

Altschul, S.F., Gish, W., Miller, W., Myers, E.W., Lipman, D.J. 1990. Basic local alignment search tool. *J. Mol. Biol.* 215(3):403–410.

Arakawa, K., Tamaki, S., Kono, N., et al. 2009. Genome Projector: Zoomable genome map with multiple views. *BMC Bioinformatics* 10:31.

Aury, J.M., Cruaud, C., Barbe, V., et al. 2008. High quality draft sequences for prokaryotic genomes using a mix of new sequencing technologies. *BMC Genomics* 9:603.

Bainbridge, M.N., Warren, R.L., He, A., Bilenky, M., Robertson, A.G., Jones, S.J. 2007. THOR: Targeted high-throughput ortholog reconstructor. *Bioinformatics* 23:2622–2624.

Batzoglou, S., Jaffe, D.B., Stanley, K., et al. 2002. ARACHNE: A whole-genome shotgun assembler. *Genome Res.* 12(1):177–189.

Butler, J., MacCallum, I., Kleber, M., et al. 2008. ALLPATHS: *De novo* assembly of whole-genome shotgun microreads. *Genome Res.* 18(5):810–820.

Chaisson, M.J.P., Brinza, D., Pevzner, P.A. 2009. *De novo* fragment assembly with short mate-paired reads: Does the read length matter? *Genome Res.* 19:336–346.

Chaisson, M.J., Pevzner, P.A. 2008. Short read fragment assembly of bacterial genomes. *Genome Res.* 18(2):324–330.

Chenna, R., Sugawara, H., Koike, T., et al. 2003. Multiple sequence alignment with the Clustal series of programs. *Nucleic Acids Res.* 31(13):3497–3500.

Chevreux, B., Pfisterer, T., Drescher, B., et al. 2004. Using the miraEST assembler for reliable and automated mRNA transcript assembly and SNP detection in sequenced ESTs. *Genome Res.* 14(6):1147–1159.

Dohm, J.C., Lottaz, C., Borodina, T., Himmelbauer, H. 2007. SHARCGS, a fast and highly accurate short-read assembly algorithm for *de novo* genomic sequencing. *Genome Res.* 17(11):1697–1706.

Finn, R.D., Clements, J., Eddy, S.R. 2011. HMMER Web server: Interactive sequence similarity searching. *Nucleic Acids Res.* 39:W29–W37.

Flicek, P., Birney, E. 2009. Sense from sequence reads: Methods for alignment and assembly. *Nat. Methods Suppl.* 6(11s):S6–S12.

Gnerrea, S., MacCalluma, I., Przybylskia, D., et al. 2011. High-quality draft assemblies of mammalian genomes from massively parallel sequence data. *Proc. Natl. Acad. Sci. USA* 108:1513–1518.

Gordon, P.M.K., Sensen, C.W. 2007. Seahawk: Moving beyond HTML in Web-based bioinformatics analysis. *BMC Bioinformatics* 8:208.

Hach, F., Hormozdiari, F., Alkan, C., et al. 2010. mrsFAST: A cache-oblivious algorithm for short-read mapping. *Nat. Methods* 7:576–577.

Harismendy, O., Ng, P.C., Strausberg, R.L., et al. 2009. Evaluation of next generation sequencing platforms for population targeted sequencing studies. *Genome Biol.* 10(3):R32.

Hernandez, D., François, P., Farinelli, L., Osterås, M., Schrenzel, J. 2008. *De novo* bacterial genome sequencing: Millions of very short reads assembled on a desktop computer. *Genome Res.* 18(5):802–809.

Huse, S., Huber, J., Morrison, H., Sogin, M.L., Welch, D.M. 2007. Accuracy and quality of massively parallel DNA pyrosequencing. *Genome Biol.* 8(7):R143.

Huson, D.H., Auch, A.F., Qi, J., Schuster, S.C. 2007. MEGAN analysis of metagenomic data. *Genome Res.* 17:377–386.

Jeck, W.R., Reinhardt, J.A., Baltrus, D.A., et al. 2007. Extending assembly of short DNA sequences to handle error. *Bioinformatics* 23(21):2942–2944.

Johnson, K.P., Walden, K.K.O., Robertson, H.M. 2010. A target restricted assembly method (TRAM) for phylogenomics. *Nature Precedings,* http://hdl.handle.net/10101/npre.2010.4612.1.

Langmead, B., Trapnell, C., Pop, M., Salzberg, S.L. 2009. Ultrafast and memory-efficient alignment of short DNA sequences to the human genome. *Genome Biol.* 10:R25.

Li, H., Durbin, R. 2009. Fast and accurate short read alignment with Burrows-Wheeler transform. *Bioinformatics* 25:1754–1760.

Li, R., Yu, C., Li, Y., et al. 2009. SOAP2: An improved ultrafast tool for short read alignment. *Bioinformatics* 25:1966–1967.

Li, R., Zhu, H., Ruan, J., et al. 2010. *De novo* assembly of human genomes with massively parallel short-read sequencing. *Genome Res.* 20(2):265–272.

Li, W., Godzik, A. 2006. Cd-hit: A fast program for clustering and comparing large sets of protein or nucleotide sequences. *Bioinformatics* 22:1658–1659.

Li, Z., Chen, Y., Mu, D., et al. 2011. Comparison of the two major classes of assembly algorithms: Overlap–layout–consensus and de-Bruijn-graph. *Brief. Funct. Genomics* 11(1):25–37.

Lin, H., Zhang, Z., Zhang, M.Q., Ma, B., Li, M. 2008. ZOOM! Zillions of oligos mapped. *Bioinformatics* 24(21):2431–2437.

Lin, Y., Li, Z., Shen, H., Zhang, L., Papasian, C.J., Deng, H.W. 2011. Comparative studies of *de novo* assembly tools for next-generation sequencing technologies. *Bioinformatics* 27(15):2031–2037.

Mardis, E.R. 2008. Next-generation DNA sequencing methods. *Annu. Rev. Genomics Hum. Genet.* 9:387–402.

Margulies, M., Egholm, M., Altman, W.E. 2005. Genome sequencing in microfabricated high-density picolitre reactors. *Nature* 437(7057):376–380.

Metzker, M.L. 2010. Sequencing technologies—The next generation. *Nat. Rev. Genet.* 11:31–46.

Miller, J.R., Delcher, A.L., Koren, S., et al. 2008. Aggressive assembly of pyrosequencing reads with mates. *Bioinformatics* 24(24):2818–2824.

Miller, J.R., Koren, S., Sutton, G. 2010. Assembly algorithms for next-generation sequencing data. *Genomics* 95:315–327.

Myers, E.W., Sutton, G.G., Delcher, A.L., et al. 2000. A whole-genome assembly of Drosophila. *Science* 287(5461):2196–2204.

Nagaraj, S.H., Gasser, R.B., Ranganathan, S. 2006. A hitchhiker's guide to expressed sequence tag (EST) analysis. *Brief. Bioinform.* 8:6–21.

Ning, Z., Cox, A.J., Mullikin, J.C. 2001. SSAHA: A fast search method for large DNA databases. *Genome Res.* 11(10):1725–1729.

Notredame, C., Higgins, D.G., Heringa, J. 2000. T-Coffee: A novel method for fast and accurate multiple sequence alignment. *J. Mol. Biol.* 302(1):205–217.

Ossowski, S., Schneeberger, K., Clark, R. M., Lanz, C., Warthmann, N., Weigel, D. 2008. Sequencing of natural strains of *Arabidopsis thaliana* with short reads. *Genome Res.* 18(12):2024–2033.

Pettersson, E., Lundeberg, J., Ahmadian, A. 2009. Generations of sequencing technologies. *Genomics* 93:105–111.

Pop, M. 2009. Genome assembly reborn: Recent computational challenges. *Brief. Bioinform.* 10(4):354–366.

Prentice, I.C. 1977. Non-metric ordination methods in ecology. *J. Ecol.* 65(1):85–94.

Pruesse, E., Quast, C., Knittel, K. 2007. SILVA: A comprehensive online resource for quality checked and aligned ribosomal RNA sequence data compatible with ARB. *Nucleic Acids Res.* 35:7188–7196.

Reinhardt, J.A., Baltrus, D.A., Nishimura, M.T., Jeck, W.R., Jones, C.D., Dangl, J.L. 2009. *De novo* assembly using low-coverage short read sequence data from the rice pathogen *Pseudomonas syringae* pv. *oryzae*. *Genome Res.* 19(2):294–305.

Robinson, J.T., Thorvaldsdóttir, H., Winckler, W., et al. 2011. Integrative genomics viewer. *Nat. Biotechnol.* 29(1):24–26.

Ruffalo, M., LaFramboise, T., Koyutürk, M. 2011. Comparative analysis of algorithms for next-generation sequencing read alignment. *Bioinformatics* 27(20):2790–2796.

Rumble, S.M., Lacroute, P., Dalca, A.V., Fiume, M., Sidow, A., Brudno, M. 2009. SHRiMP: Accurate mapping of short color-space reads. *PLoS Comput. Biol.* 5(5):e1000386.

Saeed, A.I., Sharov, V., White, J., et al. 2003. TM4: A free, open-source system for microarray data management and analysis. *Biotechniques* 34(2):374–378.

Schloss, P.D., Westcott, S.L. 2011. Assessing and improving methods used in operational taxonomic unit-based approaches for 16S rRNA gene sequence analysis. *Appl. Environ. Microbiol.* 77:3219–3226.

Schloss, P.D., Westcott, S.L., Ryabin, T., et al. 2009. Introducing mothur: Open-source, platform-independent, community-supported software for describing and comparing microbial communities. *Appl. Environ. Microbiol.* 75:7537–7541.

Shendure, J., Ji, H. 2008. Next-generation DNA sequencing. *Nat. Biotechnol.* 26(10):1135–1145.

Simpson, J.T., Wong, K., Jackman, S.D., Schein, J.E., Jones, S.J., Birol, I. 2009. ABySS: A parallel assembler for short read sequence data. *Genome Res.*19(6):1117–1123.

Skinner, M.E., Uzilov, A.V., Stein, L.D., Mungall, C.J., Holmes, I.H. 2009. JBrowse: A next-generation genome browser. *Genome Res.* 19:1630–1638.

Slater, G.S., Birney, E. 2005. Automated generation of heuristics for biological sequence comparison. *BMC Bioinformatics* 6:31.

Smith, T.F., Waterman, M.S. 1981. Identification of common molecular subsequences. *J. Mol. Biol.* 147:195–197.

Soh, J., Gordon, P.M.K., Sensen, C.W. 2012. The Bluejay genome browser. *Curr. Protoc. Bioinformatics* 10.9.1–10.9.23.

Soh, J., Gordon, P.M.K., Taschuk, M.L., et al. 2008. Bluejay 1.0: Genome browsing and comparison with rich customization provision and dynamic resource linking. *BMC Bioinformatics* 9:450.

Stanke, M., Diekhans, M., Baertsch, R., Haussler, D. 2008. Using native and syntenically mapped cDNA alignments to improve *de novo* gene finding. *Bioinformatics* 24(5):637–644.

Stanke, M., Waack, S. 2003. Gene prediction with a hidden Markov model and a new intron submodel. *Bioinformatics* 19 Suppl. 2:ii215–ii225.

Ter-Hovhannisyan, V., Lomsadze, A., Chernoff, Y.O., Borodovsky, M. 2008. Gene prediction in novel fungal genomes using an *ab initio* algorithm with unsupervised training. *Genome Res.* 18(12):1979.

Wang, Z., Gertein, M., Snyder, M. 2009. RNA-Seq: A revolutionary tool for transcriptomics. *Nat. Rev. Genet.* 10(1):57–63.

Warren, R.L., Sutton, G.G., Jones, S.J., Holt, R.A. 2007. Assembling millions of short DNA sequences using SSAKE. *Bioinformatics* 23(4):500–501.

Whittaker, R.H. 1972. Evolution and measurement of species diversity. *Taxon* 21:213–251.

Wold, S., Esbensen, K., Geladi, P. 1987. Principal component analysis. *Chemometr. Intell. Lab.* 2:37–52.

Yates, T., Okoniewski, M.J., Miller, C.J. 2008. X:Map: Annotation and visualization of genome structure for Affymetrix exon array analysis. *Nucleic Acids Res.* 36(Database issue):D780–D786.

Zerbino, D.R., Birney, E. 2008. Velvet: Algorithms for *de novo* short-read assembly using de Bruijn graphs. *Genome Res.*18(5):821–829.

Zhu, J., Sanborn, J.Z., Benz, S., et al. 2009. The UCSC Cancer Genomics Browser. *Nat. Methods* 6:239–240.

Index

For Product Safety Concerns and Information please contact our EU
representative GPSR@taylorandfrancis.com Taylor & Francis Verlag GmbH,
Kaufingerstraße 24, 80331 München, Germany

Printed and bound by CPI Group (UK) Ltd, Croydon, CR0 4YY
01/05/2025
01858332-0001